The author, Norbert Waldy, has developed since the 1980s different safety procedures for machinery manufacturing in order to increase security standards.

He has invariably focused his attention on the practical aspect and always believed that self-explanatory check lists were the key to success.

After 10 years in the pharmaceutical industry in which he had the occasion to qualify, validate and monitor complex pharmaceutical facilities as Head of Department, he decided to transmit his knowledge with this book.

The author is an accredited auditor of Quality Management System ISO 9001 and OHSAS 18001 machine manufacturing since 15 years. He has certified several small and medium-sized enterprises in the course of his function.

Norbert.waldy@hotmail.de

Translator
Rose Mary Herren-Gleeson
International experience
Translation services
German-English
German-French
r.herren-gleeson.translations@bluewin.ch

www.tredition.de

© 2013 Norbert Waldy - Rose Mary Herren-Gleeson
Edition: 2013

Publisher: tredition GmbH, Hamburg
ISBN: ISBN: 978-3-8495-7292-1
Printed in Germany

Bibliografische Information der Deutschen Nationalbibliothek:
Die Deutsche Nationalbibliothek verzeichnet diese Publikation in
der Deutschen Nationalbibliografie; detaillierte bibliografische Da-
ten sind im Internet über http://dnb.d-nb.de abrufbar.

Norbert Waldy
Rose Mary Herren-Gleeson

# CE marking
# for machinery and
# partly completed machinery
# from practitioner to
# practitioner

# Table of contents

# Introduction

The CE marking is a Passport for Machines in Europe.

With the CE marking, a manufacturer declares having respected and applied the corresponding harmonised standard guidelines for his product.

At the same time, he has successfully carried out the "adapted" conformity assessment".

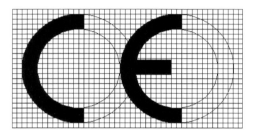

This book should allow the practitioner to carry out the required conformity assessment on his own without much effort.

Besides this, I have included practical examples like: risk assessment, conformity assessment, CE-declaration of conformity, acceptance protocol, processing sequences and several other tools.

# 1. The CE-Delegate

The Task and competence of a CE-Delegate should be regulated by a (his) job description

## 1.1. Tasks of a CE-Delegate

- Development of the conformity assessment
- Creation of the risk assessment
- Staff training concerning CE marking
- Collaboration with the authorities
- Contact person for the authorities
- Determination of the examinations
- Backing of the product from design until delivery
- Determination of the retention period of documents
- Creation or help with technical documents
- Issuing of EC-Declaration of conformity
- Managing of standards and guidelines
- Make sure that the applied standards and guidelines are up to date
- Direct reporting to the direction or the manager
- Technical consulting within the company in all questions regarding the CE marking
- Keeping the EC-Declaration of conformity and technical documentation ready for the national surveillance authorities

## 1.2. Competence of the CE-Delegate

- Imposing a delivery stop for nonconforming products
- Imposing a delivery stop for nonconforming technical file
- Has the authority to sign an EC-Declaration of conformity
- Has the authority to sign any technical file
- Has the right to regular further training

- Reporting to authorities / Obligation of self-denunciation in case of gross carelessness

## 1.3.    Role of the CE-Delegate

As a specialist, the CE-Delegate should report to the company management or to the General Manager and keep them regularly informed.

## 1.4.    Organization chart

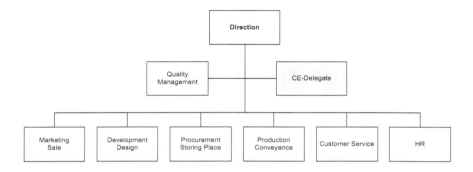

# 2. Authorised representative, technical file

The Machinery Directive 2006/42/EC requires the designation of an authorised representative for the technical file related to an EC-Declaration of conformity:

- Name and address of the person authorised to compile the corresponding technical documents and,
- Address where the controlling authorities may request the documentation.

Definition of authorised representative

- authorised representative' means any natural or legal person established in the Community who has received a written mandate from the manufacturer to perform on his behalf all or part of the obligations and formalities connected with this Directive.

## 2.1. Tasks, Responsibility of the authorised representative

- The authorised representative is the contact person for the Authorities
- The authorised representative is not responsible for the exactitude of the technical file
- The authorised representative is mentioned in the declaration of conformity
- The manufacturer is responsible for the exactitude of the technical file.
- The manufacturer (see definition) and the authorised representative can be different persons with different addresses
- The authorised representative must be a resident of the EU
- It can be a natural or a juristic person

Definition of manufacturer:

- 'manufacturer' means any natural or legal person who designs and/or manufactures machinery or partly completed machinery covered by this Directive and is responsible for the conformity of the machinery or the partly completed machinery with this Directive with a view to its being placed on the market, under his own name or trademark or for his own use. In the absence of a manufacturer as defined above, any natural or legal person who places on the market or puts into service machinery or partly completed machinery covered by this Directive shall be considered a manufacturer;

# 3. The way to CE marking

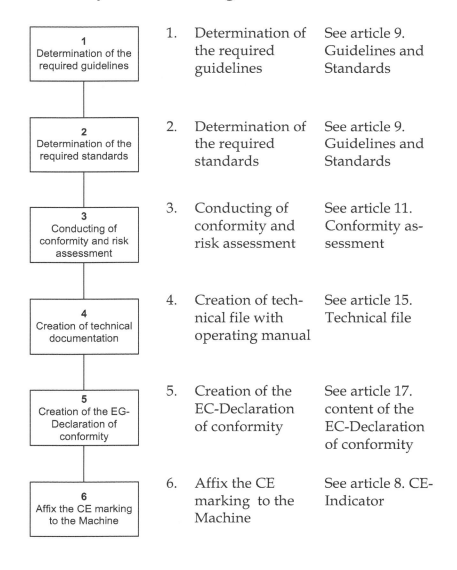

| | | |
|---|---|---|
| 1. | Determination of the required guidelines | See article 9. Guidelines and Standards |
| 2. | Determination of the required standards | See article 9. Guidelines and Standards |
| 3. | Conducting of conformity and risk assessment | See article 11. Conformity assessment |
| 4. | Creation of technical file with operating manual | See article 15. Technical file |
| 5. | Creation of the EC-Declaration of conformity | See article 17. content of the EC-Declaration of conformity |
| 6. | Affix the CE marking to the Machine | See article 8. CE-Indicator |

# 4. The CE Procedure

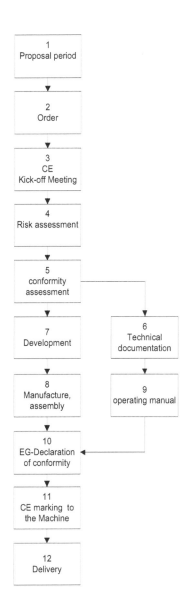

1. Proposal period

2. Order

3. Kick-off Meeting

4. Creation of risk assessment

5. Development of the conformity assessment

6. Preparation of the technical file

7. Development and construction of the machine, taking account of points 4 and 5

8. Manufacture, assembly and testing of the machine

9. Creation of the operating manual

10. Issue of the EC-Declaration of conformity

11. Affix the CE marking to the Machine

12. Delivery of the machine

# 5. The Machinery Directive 2006/42/EC

The CE-Declaration of conformity: The "Passport" for introduction and placing machinery on the European market

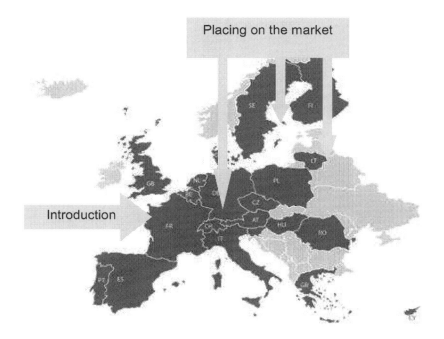

## 5.1. Definition of Machinery

Excerpt from the Machinery Directive 2006/42/EC:

- an assembly, fitted with or intended to be fitted with a drive system other than directly applied human or animal effort, consisting of linked parts or components, at least one of which moves, and which are joined together for a specific application

## 5.2 Definition of partly completed machinery

Excerpt aus the Machinery Directive 2006/42/EC:
- partly completed machinery' means an assembly which is almost machinery but which cannot in itself perform a specific application. A drive system is partly completed machinery. Partly completed machinery is only intended to be incorporated into or assembled with other machinery or other partly completed machinery or equipment, thereby forming machinery to which this Directive applies.

## 5.3 Scope of the Machinery Directive

The Machinery Directive 2006/42/EC applies to following products:
- machinery
- interchangeable equipment
- safety components
- lifting accessories
- chains, ropes and webbing
- removable mechanical transmission devices
- partly completed machinery

## 5.4 Excluded from the scope

The following are excluded from the scope of this Directive:
- safety components intended to be used as spare parts to replace identical components and supplied by the manufacturer of the original machinery;
- specific equipment for use in fairgrounds and/or amusement parks;
- machinery specially designed or put into service for nuclear purposes which, in the event of failure, may result in an emission of radioactivity;
- weapons, including firearms

The following means of transport:

- agricultural and forestry tractors for the risks covered by Directive 2003/37/EC, with the exclusion of machinery mounted on these vehicles,
- motor vehicles and their trailers covered by Council Directive 70/156/EEC of 6 February 1970 on the approximation of the laws of the Member States relating to the type-approval of motor vehicles and their trailers (1), with the exclusion of machinery mounted on these vehicles,
- vehicles covered by Directive 2002/24/EC of the European Parliament and of the Council of 18 March 2002 relating to the type-approval of two or three-wheel motor vehicles (2), with the exclusion of machinery mounted on these vehicles,
- motor vehicles exclusively intended for competition, and
- means of transport by air, on water and on rail networks with the exclusion of machinery mounted on these means of transport;
- seagoing vessels and mobile offshore units and machinery installed on board such vessels and/or units
- machinery specially designed and constructed for military or police purposes
- machinery specially designed and constructed for research purposes for temporary use in laboratories
- mine winding gear
- machinery intended to move performers during artistic performances
- electrical and electronic products falling within the following areas, insofar as they are covered by Council Directive 73/23/EEC of 19 February 1973 on the harmonisation of the laws of Member States relating to

electrical equipment designed for use within certain voltage limits (3):

- household appliances intended for domestic use,
- audio and video equipment,
- information technology equipment,
- ordinary office machinery,
- low-voltage switchgear and control gear,
- electric motors;

The following types of high-voltage electrical equipment:

- switch gear and control gear,
- transformers.

## 5.5    Penalties

Member States shall lay down the rules on penalties applicable to infringements of the national provisions adopted pursuant to this Directive.

The penalties provided for must be effective, proportionate and dissuasive.

# 6. Which procedure applies

**IMPORTANT:** according to the Machinery Directive 2006/42/EC the manufacturer or his authorised representative shall apply one of the procedures for assessment of conformity described in paragraphs 2, 3 and 4, in order to certify the conformity of machinery.

## 6.1   Product classification

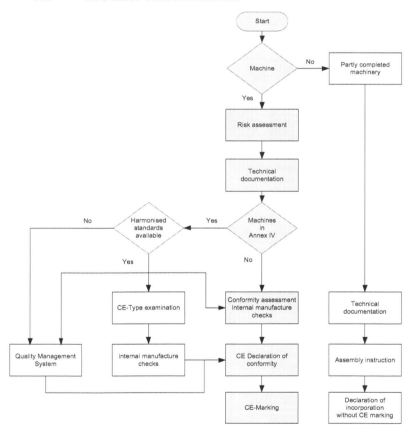

## 6.2    Conformity assessment and modules

Requirements of each procedure are exactly described in the Directive 768/2008/EC annex II, procedure of conformity assessment Modules A to H, see article 7  „Modules A to H "

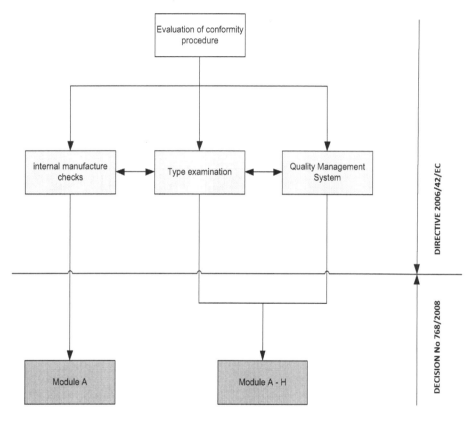

## 6.3    Internal manufacture checks

Where the machinery is not referred to in Annex IV, the manufacturer or his authorised representative shall apply the procedure for assessment of conformity with internal checks on the manufacture of machinery provided for in Annex VIII

.

**List of categories** of machines **from Annex IV** to which the type examination must be applied

1.    Circular saws (single- or multi-blade) for working with wood and material with similar physical characteristics or for working with meat and material with similar physical characteristics, of the following types:
   - sawing machinery with fixed blade(s) during cutting, having a fixed bed or support with manual feed of the workpiece or with a demountable power feed;
   - sawing machinery with fixed blade(s) during cutting, having a manually operated reciprocating saw-bench or carriage;
   - sawing machinery with fixed blade(s) during cutting, having a built-in mechanical feed device for the workpieces, with manual loading and/or unloading;

2.    Hand-fed surface planing machinery for woodworking.

3.    Thicknessers for one-side dressing having a built-in mechanical feed device, with manual loading and/or unloading for woodworking.

4.    Band-saws with manual loading and/or unloading for working with wood and material with similar physical characteristics or for working with meat and material with similar physical characteristics, of the following types:

- sawing machinery with fixed blade(s) during cutting, having a fixed or reciprocating-movement bed or support for the workpiece;
- sawing machinery with blade(s) assembled on a carriage with reciprocating motion.

5. Combined machinery of the types referred to in points 1 to 4 and in point 7 for working with wood and material with similar physical characteristics.

6. Hand-fed tenoning machinery with several tool holthes for woodworking.

7. Hand-fed vertical spindle moulding machinery for working with wood and material with similar physical characteristics.

8. Portable chainsaws for woodworking.

9. Presses, including press-brakes, for the cold working of metals, with manual loading and/or unloading, whose moving working parts may have a travel exceeding 6 mm and a speed exceeding 30 mm/s.

10. Injection or compression plastics-moulding machinery with manual loading or unloading.

11. Injection or compression rubber-moulding machinery with manual loading or unloading.

12. Machinery for underground working of the following types:
- locomotives and brake-vans;
- hydraulic-powered roof supports.

13. Manually loaded trucks for the collection of household refuse incorporating a compression mechanism.

14. Removable mechanical transmission devices including their guards.

15. Guards for removable mechanical transmission devices.

16. Vehicle maintenance lifts.

17. Devices for the lifting of persons or of persons and goods involving a hazard of falling from a vertical height of more than three metres.

18. Portable cartridge-operated fixing and other impact machinery.

19. Protective devices designed to detect the presence of persons.

20. Power-operated interlocking moving guards designed to be used as safeguards in machinery referred to in points 9, 10 and 11.

21. Logic units to ensure safety functions.

22. Roll-over protective structures (ROPS).

23. Falling-object protective structures (FOPS).

## 6.4    Type examination

The manufacturer or his authorised representative is to perform the following procedure, if the machine is included in the above mentioned list,

- Internal checks and EC type examination during manufacturing
- Extensive Quality Management procedure

## 6.5    Extensive Quality Management procedure

Should the machine not be included in the above mentioned list, the corresponding security and health requirements not be respected, or harmonised standards not exist for the machine concerned, the manufacturer or his authorised representative are to follow the succeding procedure:

- EC type examination and internal checks while manufacturing
- Extensive Quality Management procedure

# 7.   Modules A to H

In order to affix a CE marking to a machine, a conformity assessment must be carried out.

Several modules are available to do this. They are explicitly described in the Directive 768/2008/EC.

## 7.1   Module A

Internal checks

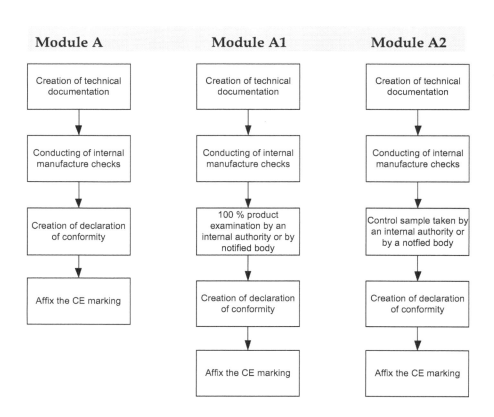

## 7.2    Module B

The EC type examination

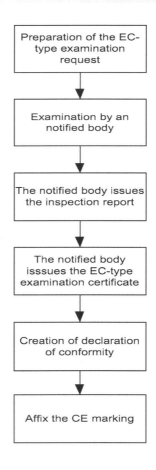

## Module B

Preparation of the EC-type examination request

Examination by an notified body

The notified body issues the inspection report

The notified body isssues the EC-type examination certificate

Creation of declaration of conformity

Affix the CE marking

## 7.3    Module C

Design conformity based on an internal checks

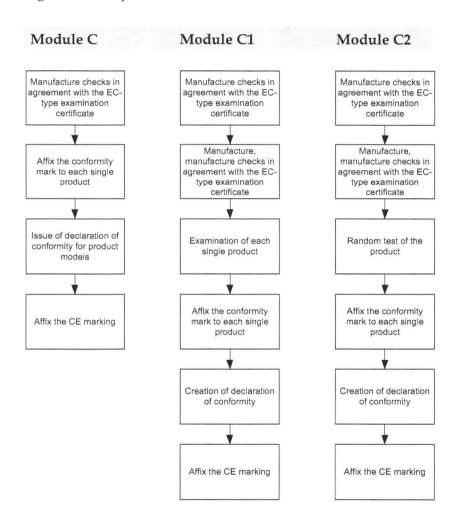

## 7.4    Module D

Design conformity based on Quality Management concerning the manufacturing procedure

**Module D**                    **Module D1**

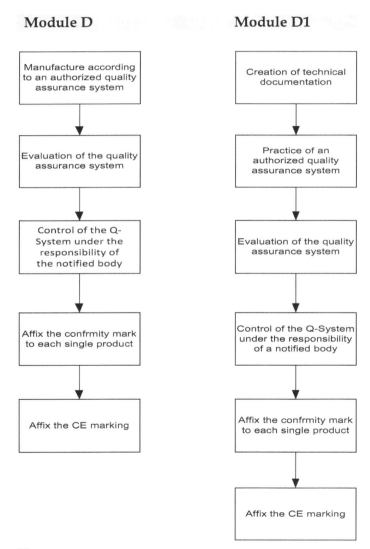

## 7.5 Module E

Design conformity based on Quality Management concerning the product

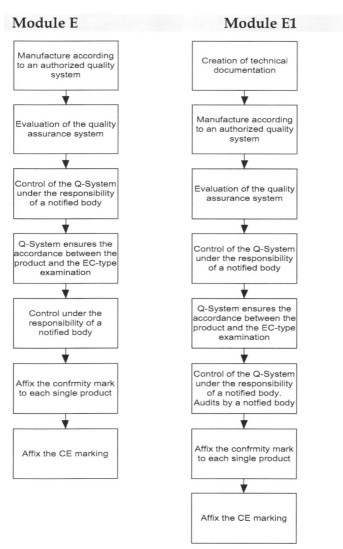

## 7.6    Module F

Design conformity based on a product examination

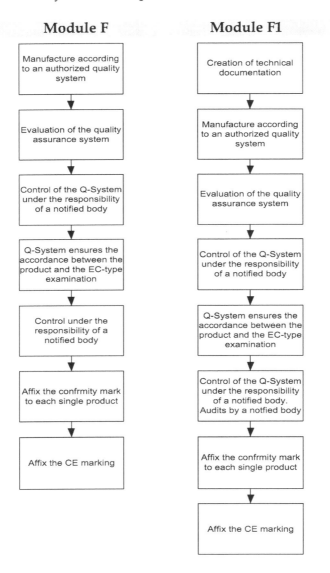

## 7.7    Module G

Conformity based on individual examination

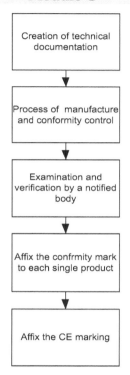

**Module G**

Creation of technical documentation

↓

Process of manufacture and conformity control

↓

Examination and verification by a notified body

↓

Affix the confrmity mark to each single product

↓

Affix the CE marking

## 7.8   Module H

Conformity based on an extensive Quality Management procedure

**Module H**

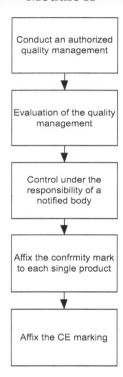

# 8.    CE-Marking

## 8.1    Compulsory CE-Marking

Excerpt from the Directive 768/2008/EC:
- The CE marking is compulsory for any product subject to it before putting on the market or into circulation
- The CE marking, indicating the conformity of a product, is the visible consequence of a whole process comprising conformity assessment in a broad sense
- It is crucial to make clear to both manufacturers and users that by affixing the CE marking to a product the manufacturer declares that the product is in conformity with all applicable requirements and that he takes full responsibility therefor.

Miscellaneous
- The CE marking  is not for commercial use
- This identification is compulsory for all products, whether manufactured in a member country or not
- This obligation is also applicable for new or second-hand products imported from other countries

## 8.2    CE marking

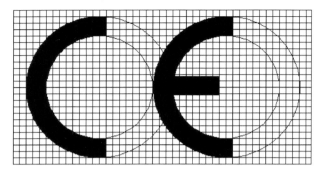

The CE marking must be affixed by the manufacturer or his authorised representative in the community.

A manufacturer declares with the CE marking having applied the corresponding harmonised standards, guidelines and regulations, and carried out the related conformity assessment as well.

The CE marking is often called the „Passport" to the domestic market.

CE means „Communauté Européenne" (European community).

The CE marking must be affixed in immediate vicinity of the name of the manufacturer or his authorised representative, using the same technique.

The CE marking is neither a quality seal nor a hallmark.


Excerpt from the Directive 768/2008/EC:
*   The CE marking shall be affixed visibly, legibly and indelibly to the product or to its data plate. Where that is not pos-

sible or not warranted on account of the nature of the product, it shall be affixed to the packaging and to the accompanying documents, where the legislation concerned provides for such documents.

- The CE marking should be the only marking of conformity indicating that a product is in conformity with Community harmonisation legislation. However, other markings may be used as long as they contribute to the improvement of consumer protection and are not covered by Community harmonisation.

Requirement of the Machinery Directive 2006/42/EC: the font should be at least 5 mm size

Marks, signs or superscripts whose meaning or shape could be misunderstood or taken for a CE marking may not be affixed to machinery.

Any other mark whose visibility, legibility and meaning does impair the CE marking legibility can be affixed

## 8.3    Marking of machinery

Following indications must be legible and permanently affixed to each machine:

- company name and full address of the manufacturer and, if applicable, the name of his authorised representative;
- label of the machine;
- CE marking;
- class or type description;
- serial number, if applicable;
- year of manufacture, i.e. the year in which the manufacture procedure was finished.

It is not allowed to antedate or postdate the manufacture year of the machine when affixing the CE marking.

If the machine is designed and manufactured to be used in an explosive area, an indication must be made in this sense.

Depending on its type, all relevant information for a safe utilization must be included on the machine as well.

Any machinery needing lifting gear for its use must have its weight shown in a permanent and clearly readable way.

## 8.4 Non-conformity of marking

Member States shall consider the following marking not to conform:

- the affixing of the CE marking pursuant to this Directive
- on products not covered by this Directive.
- the absence of the CE marking and/or the absence of the EC declaration of conformity for machinery
- the affixing on machinery of a marking, other than the CE
- marking, which is prohibited under Article 16(3.

Where a Member State ascertains that marking does not conform to the relevant provisions of this Directive, the manufacturer or his authorised representative shall be obliged to make the product conform and to put an end to the infringement under conditions fixed by that Member State.
Where non-conformity persists, the Member State shall take all appropriate measures to restrict or prohibit the placing on the market of the product in question or to ensure that it is withdrawn from the market.

## 8.5 Sales literature

Sales literature describing the machinery must not contradict the instructions as regards health and safety aspects.

Sales literature describing the performance characteristics of machinery must contain the same information on emissions as is contained in the instructions

.

## 8.6 Machinery trades and demonstrations

At trade fairs, exhibitions, demonstrations, and such like, Member States shall not prevent the showing of machinery or partly completed machinery which does not conform to this Directive, provided that a visible sign clearly indicates that it does not conform and that it shall not be made available until it has been brought into conformity.

Furthermore, during demonstrations of such nonconforming machinery or partly completed machinery, adequate safety measures shall be taken to ensure the protection of persons.

# 9. Guidelines and Standards

All valid guidelines and standards are published on the homepage of the European Union „Nando".

Guidelines are law and must be respected accordingly; they are made expressed in concrete terms in corresponding harmonised standards.

# 10.    The ISO 12100

Excerpt from the Machinery Directive 2006/42/EC:

- The manufacturer of machinery or his authorised representative must ensure that a risk assessment is carried out in order to determine the health and safety requirements which apply to the machinery. The machinery must then be designed and constructed taking into account the results of the risk assessment

The risk assessment is undertaken before the machine is designed and constructed (Design period).

Results of the risk assessment may be used to establish the acceptance protocol (this is regularly done in the Pharmaceutical industry, FAT, SAT).

## 10.1    Strategy for the risk assessment / reduction

To carry out a risk assessment and achieve a risk reduction, the following steps must be taken:

- Determine the limits of the machinery, which include the intended use and any reasonable foreseeable misuse thereof,
- identify the hazards that can be generated by the machinery and the associated hazardous situations,
- estimate the risks, taking into account the severity of the possible injury or damage to health and the probability of its occurrence,
- evaluate the risks, with a view to determining whether risk reduction is required, in accordance with the objective of this Directive,
- eliminate the hazards or reduce the risks associated with these hazards by application of protective measures,

The ISO 12100:2010 is a good tool to inquire about machinery hazards. The information obtained can be integrated and suits as hazard catalogue for the risk assessment.

ISO 12100:2010 establishes guidelines for the risk assessment and risk reduction to support manufacturers ensuring a safe machinery construction.

## 10.2    Mechanical hazards

- acceleration, deceleration
- angular parts
- approach of a moving element to a fixed part;
- cutting parts
- elastic elements
- falling objects
- gravity
- height from the ground
- high pressure
- instability
- kinetic energy
- machinery mobility
- moving elements
- rotating elements
- rough, slippery surface
- sharp edges
- stored energy
- vacuum
- being run over
- being thrown
- crushing
- cutting or severing
- drawing-in or trapping
- entanglement
- friction or abrasion

- impact
- injection
- shearing
- slipping, tripping and falling
- stabbing or puncture
- suffocation
- Contact with rough surfaces
- Contact with sharp edges and corners, protruding parts
- Contact with moving parts
- Contact with rotating open ends
- Falling or ejection of objects
- Loss of stability
- Break-up during operation
- Displacement of moving elements
- Projection of high pressure fluids
- Uncontrolled movements
- other mechanical hazards

## 10.3    Electrical hazards

- arc
- electromagnetic phenomena
- electrostatic phenomena
- live parts
- not enough distance to live parts under high voltage
- overload
- parts which have become live under fault conditions
- short-circuit
- thermal radiation
- burn
- chemical effects
- effects on medical implants
- electrocution
- falling, being thrown

- fire
- projection of molten particles
- shock
- direct contact
- disruptive discharge
- electric arc
- indirect contact
- short-circuit

## 10.4    Thermal Hazards

- explosion
- flame
- objects or materials with a high
  or low temperature
- radiation from heat sources
- burn
- dehydration
- discomfort
- frostbite
- injuries by the radiation of heat sources
- scald

## 10.5    Noise hazards

- cavitation phenomena
- exhausting system
- gas leaking at high speed
- manufacturing process (stamping, cutting, etc.);
- moving parts
- scraping surfaces
- unbalanced rotating parts
- whistling pneumatics
- worn parts
- discomfort

- loss of awareness
- loss of balance
- permanent hearing loss
- stress
- tinnitus
- tiredness

any other (for example, mechanical, electrical) as a consequence of an interference with speech communication or with acoustic signals

## 10.6    Vibration hazards

- cavitation phenomena
- misalignment of moving parts
- mobile equipment
- scraping surfaces
- unbalanced rotating parts
- vibrating equipment
- worn parts
- discomfort
- low-back morbidity
- neurological disorder
- osteo-articular disorder
- trauma of the spine
- vascular disorder

## 10.7    Radiation hazards

- ionizing radiation source
- low frequency electromagnetic optical radiation (infrared, visible and ultraviolet), including laser
- radio frequency electromagnetic radiation
- burn
- damage to eyes and skin
- effects on reproductive capability
- mutation
- headache, insomnia, etc.

## 10.8    Material/substance hazards

- aerosol
- biological and microbiological (viral or bacterial) agent
- combustible
- dust
- explosive

- fibre
- flammable
- fluid
- fume
- gas
- mist
- oxidizer
- breathing difficulties, suffocation
- cancer
- corrosion
- effects on reproductive capability
- explosion
- fire
- infection
- mutation
- poisoning
- sensitization

## 10.9   Ergonomic hazards

- access
- design or location of indicators and visual displays units
- design, location or identification of control devices
- effort
- flicker, dazzling, shadow, stroboscopic effect
- local lighting
- mental overload/underload
- posture
- repetitive activity
- visibility
- discomfort
- fatigue
- musculoskeletal disorder
- stress

- any other (for example, mechanical, electrical) as a consequence of a human error

## 10.10   Hazards associated with the environment

- dust and fog
- electromagnetic disturbance
- lightning
- moisture
- pollution
- snow
- temperature
- water
- wind
- lack of oxygen
- burn
- slight disease;
- slipping, falling
- suffocation
- any other as a consequence of the effect caused by the sources of the hazards on the machine or parts of the machine

## 10.11   Combination of hazards

- for example, repetitive activity +effort + high environmental temperature
- dehydration
- loss of awareness
- heat stroke
- dropping or ejection of a moving part of the machine or of a workpiece clamped by the machine
- failure to stop moving parts
- machine action resulting from inhibition (defeating or failure) of protective devices

- uncontrolled movements (including speed change)
- unintended/unexpected start-up
- other hazardous events due to failure(s) or poor design of the control system

# 11. Conformity assessment

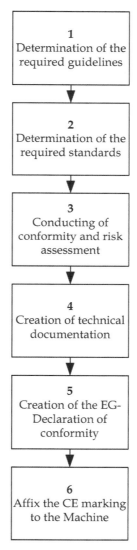

A conformity assessment includes the following steps:
- product description (application area, user / operator and limits of the machine are to be considered);
- determination of corresponding harmonized legislation;
- determination of applicable standards
- carrying out a risk assessment
- creation of technical documentation
- issuing of the CE declaration of conformity
- affix the CE marking

In order to be allowed to affix a CE marking to machinery, a conformity assessment has to be performed. See article 4 of the Directive 768/2008/EC. The following modules are available to help you carry out this procedure: see paragraph 7, Modules A to H.

## 11.1    Limits of the Machine

The limits of a machine and its durability are to be determined from the beginning of the conformity assessment.

Examples of limits:
- application limits
- spacial limits
- time limits
- environmental temperature
- operating place
- cleaning
- maximal performance of the machine
- minimal performance of the machine
- education of end-user

## 11.2    Application limits

They include the intended use and the reasonably foreseeable misuse.

Aspects to be considered are as follows:
- all different operating conditions by the end-user, including those becoming necessary after misuse of the machine,
- application area (Industry, trade, household,...),
- user (different gender, age, left or right-handed or with limited physical skills, visually handicapped or hearing-impaired, body size, force, etc.)
- necessary qualification level, experience or skills of users including operating staff.
- maintenance staff or technicians
- apprentices and trainee
- public
- children
- any other person which could be exposed to a reasonably foreseeable hazard related to the machine
- any person who is most probably aware of the specific hazards, like operating personnel
- any person unaware of the specific hazards and yet knowing the security procedures related to the location, allowed roads, etc., like public administration staff
- any person neither aware of the specific hazards nor of the security procedures related to the location
- any visitor or public

## 11.3   Spacial limits

The spacial aspects to be considered are:
- scope;
- free scope for persons needing access to the machine like maintenance or operating staff;
- interaction between person and machine, like for example the interface "person/machine" and „machine/power supply"

## 11.4    Time limits

The time limits to be considered are:
- durability of the machine and/or its components (like tools, attrition parts, electrical devices) considering their designated use and any reasonably foreseeable misuse
- recommendable maintenance intervals

## 11.5    Further limits

Examples of further limits are:
- characteristics of the materials to be processed
- maintenance
- cleaning
- degree of cleanliness
- ambient temperature (lowest and highest)
- operating place:
    - indoors
    - outdoors
    - in dry weather conditions
    - in wet weather conditions
    - under direct solar exposure
    - in dust
    - in snow, foggy, or windy conditions, with limited visibility

## 11.6    Determination of tasks

When determining tasks, the whole life-cycle of the machine should be considered.

Example of possible tasks:
- installing
- checking
- starting
- programme

- conversion
- training
- loading
- unloading
- stop of the machine
- stop in emergency cases
- fixing after a jam
- restart after an unforeseen shut-down
- troubleshooting and elimination of errors
- intervention in the procedure
- cleaning
- preventive maintenance
- calibration
- maintenance to eliminate errors

# 12. Conformity assessment for punching machine

In this chapter you shall see an example of the procedure for a punching machine. The author accepts no liability for it.

## 12.1 Information about client and machine

| | |
|---|---|
| Designation of the machine | *Punching machine* |
| Type | *ST1* |
| Year of manufacture | *2013* |
| Serial number | *132654789EZB* |
| Client | *Stanztechnik AG, Stanzerstrasse 1, 123456 Zurich* |
| Manufacturer | *Mustermann AG Musterstr. 5 123456 Zurich* |
| Project manager | *René Muster* |
| EC representative | *Hans Muster* |
| in charge of the technical documentation | *Hans Muster* |

## 12.2    Definition of the corresponding guidelines

Indicate the corresponding guidelines in following table

| Guide-line | Designa-tion | applica-ble | Short description of the guide-line |
|---|---|---|---|
| 06/42 | Machine Guideline | ☑ yes<br>☐ no | Machine:<br>An assembly, fitted with or intended to be fitted with a drive system other than directly applied human or animal effort, consisting of linked parts or components, at least one of which moves, and which are joined together for a specific application |
| | | ☐ yes<br>☑ no | Partly completed machine:<br>An assembly which is almost machinery but which cannot in itself perform a specific application. A drive system is partly completed machinery. Partly completed machinery is only intended to be incorporated into or assembled with other machinery or other partly completed machinery or equipment, thereby forming machinery to which this Directive applies. |

| Guide-line | Designa-tion | applica-ble | Short description of the guide-line |
|---|---|---|---|
| 94/9 | ATEX | ❏ yes<br>☑ no | Usage of the Machine in an explosive surrounding<br><br>Determination of areas:<br>❏ area 0<br>❏ area 1<br>❏ area 2 |
| 93/42 | Medical device | ❏ yes<br>☑ no | Medical device: it's an instrument, apparatus, in vitro reagent, software, substance or similar, including any item from the manufacturer, produced:<br>• To diagnose, prevent or treat a disease or other conditions<br><br>and not achieving its purposes through chemical action within or on the body. |
|  | Pressure equipment guideline | ❏ yes<br>☑ no | Pressure vessel with a maximum authorized operating pressure of 0,5 bar<br><br>Determination of categories:<br>❏ Category I<br>❏ Category II<br>❏ Category III<br>❏ Category IV |

| Guide-line | Designa-tion | applica-ble | Short description of the guide-line |
|---|---|---|---|
| 04/108 | EMC | ☐ yes<br>☑ no | EMC aims to ensure that equipment items or systems shall not interfere with or prevent each other's correct operation through spurious emission and absorption of EMI (electromagnetic interference) |
| 98/79 | In vitro diagnostics | ☐ yes<br>☑ no | 'in vitro diagnostic medical device' means any medical device which is a reagent, reagent product, calibrator, control material, kit, instrument, apparatus, equipment, or system, whether used alone or in combination, intended by the manufacturer to be used in vitro for the examination of specimens, including blood and tissue donations, derived from the human body, solely or principally for the purpose of providing information concerning:<br>• a physiological or pathological state,<br>• a congenital abnormality<br>• to determine the safety and compatibility with potential recipients<br>• to monitor therapeutic measures<br>The manufacturer bears responsibility for deciding if a product is an in-vitro diagnostic device, and for determining its application. |

| Guide-line | Designa-tion | applica-ble | Short description of the guide-line |
|---|---|---|---|
| 06/95 | Low volt-age guide-line | ❑ yes<br>☑ no | The directive covers electrical equipment with a voltage at input or output terminals be-tween 50 and 1000 V for alter-nating current (AC) or between 75 and 1500 V for direct current (DC) except equipment and areas mentioned by Annex II. |

The corresponding guidelines are registered in the EC-Declaration of conformity.

## 12.3    Definition of the corresponding standards

Following standards are to be considered in the manufacture of machinery:

| Norm | Description the Norm | Required |
|------|----------------------|----------|
| ISO 4414 | Pneumatic fluid power - General rules and safety requirements for systems and their components | ☑ Yes<br>❑ No |
| ISO 4413 | Hydraulic fluid power - General rules and safety requirements for systems and their components | ☑ Yes<br>❑ No |
| ISO 14120 | Guards - General requirements for the design and construction of fixed and movable guards | ☑ Yes<br>❑ No |
| ISO 14119 | Interlocking devices associated with guards. | ☑ Yes<br>❑ No |
| ISO 14118 | Prevention of unexpected start-up | ❑ Yes<br>☑ No |
| ISO 13857 | Safety of machinery - Safety distances to prevent hazard zones being reached by upper and lower limbs | ☑ Yes<br>❑ No |
| ISO 13855 | Safety of machinery - Positioning of safeguards with respect to the approach speeds of parts of the human body | ☑ Yes<br>❑ No |
| ISO 13854 | Safety distances to prevent danger zones being reached by the upper limbs | ❑ Yes<br>☑ No |
| ISO 13851 | Two-hand control devices. | ☑ Yes<br>❑ No |

| Norm | Description the Norm | Required |
|---|---|---|
| EN 60204-1 | Safety of machinery - Electrical equipment of machines - Part 1: General requirements | ☑ Yes ☐ No |
| EN 60079 | Explosive atmospheres | ☐ Yes ☑ No |
| EN 13850 | Emergency stop - Principles for design | ☑ Yes ☐ No |
| EN 13849-1 | Safety-related parts of control systems | ☑ Yes ☐ No |
| EN 12100 | General principles for design - Risk assessment and risk reduction | ☑ Yes ☐ No |

The corresponding guidelines are registered in EC-Declaration of conformity

## 12.4 Application limits of a machine

Possible operating modes

☐ Automatic

☑ Manual

Are necessary intervantions in case of hazardous malfunctioning?

☐ Yes - ☑ No

Application field of the Machine

☑ Manufacture

☑ Trade

Operators of the machine

☑ of different gender

☑ of different age

☑ left-, right-handed

☐ with limited physical skills

Users' qualification level related to education and experience is to be considered

User:          ☑   qualified   ☑   unqualified   ☑   semiskilled
               ☑ apprentice ❑ general public
Maintenance:   ☑   qualified   ☑   unqualified   ☑   semiskilled
               ☑ apprentice ❑ general public
Technician:    ☑   qualified   ❑   unqualified   ❑   semiskilled
               ❑ apprentice ❑ general public

Are there any hazards related to the machine which can be generated by:

☑ any person perfectly aware of the hazards, like maintenance staff working at neighbouring machines

❑ any person unaware of the specific hazards and yet knowing the security procedures related to the location, allowed roads, etc., like public administration staff

❑ any person probably unaware of the hazards related to the machine like children and public

## 12.5   Spacial limits

Following aspects are to be considered:

☑   scope

☑   free scope for persons needing access to the machine like maintenance or operating staff

❑   interaction between person and machine, like for example the interface "person/machine"

❑   interaction „machine/power supply"

❑   design:                    _____

❑   layout:                    _____

❑   floor loading plan         _____

❑   R&I schema                 _____

## 12.6    Time limits

The following aspects are to be considered:

- ☑ durability of the machine and/or its components (like tools, attrition parts, electrical devices) considering their intended use and any reasonably foreseeable misuse

Durability in years: *10*

Durability in hours: -

Recommended maintenance periodicity:
- ☑ annual maintenance
- ☑ monthly maintenance
- ❑ weekly maintenance
- ☑ daily maintenance

## 12.7    Other limits

Some examples are:
- ☑ characteristics of the materials to be processed
- ❑ cleaning and maintenance
- ☑ recommended lowest/highest temperature: 5° to 25°C

Operating place:
- ☑ indoors
- ❑ outdoors
- ❑ in dry weather conditions
- ❑ under direct solar exposure
- ❑ dust and wetness resistance
- ❑ in snow, ❑ windy, ❑ rainy, ❑ foggy conditions

## 12.8 Stage of life and operating mode of a machine

| Operating modes | Operator | Installer | Maintenance staff | Visitors, children, others | Client | Carrier | Dumping enterprise | Specialised company | Stanztechnik AG |
|---|---|---|---|---|---|---|---|---|---|
| **1)    Transport and putting into service** | | | | | | | | | |
| Packaging, Transportation | | | | | | X | | | |
| Load and unload | | | | | | X | | | |
| Assembly | | | | | | | | | X |
| Putting into service | X | X | | | | | | | X |
| **2)    Installing, operating, trouble-shooting, maintenance and cleaning** | | | | | | | | | |
| Installing | X | X | | | | | | | |
| Programming | X | X | | | | | | | |
| Modifying | | X | | | | | | | |
| Standard operation | X | | | | | | | | |
| Special operation | | X | | | | | | | |
| Production failure | | X | X | | | | | | |
| Machine malfunction | | X | X | | | | | | |
| Trouble-shooting | X | X | X | | | | | | X |
| Maintenance | X | X | X | | | | | | X |
| Cleaning | X | X | X | | | | | | |
| **3)    Decommissioning, disassembly and dumping** | | | | | | | | | |
| Disassembly | | | | | | | X | | |
| removal, dumping | | | | | | | X | | |

## 12.9 Content of the technical file

| Technical Documentation | Required |
|---|---|
| Conformity assessment | ☑ Yes ❑ No |
| Risk assessment | ☑ Yes ❑ No |
| R&I Schema | ❑ Yes ☑ No |
| Layout | ☑ Yes ❑ No |
| Floor loading plan | ☑ Yes ❑ No |
| Circuit layout | ☑ Yes ❑ No |
| Detail drawing | ☑ Yes ❑ No |
| Layout chart | ☑ Yes ❑ No |
| Pneumatics schema | ☑ Yes ❑ No |
| Hydraulics schema | ☑ Yes ❑ No |
| Calculations and experimental results | ☑ Yes ❑ No |
| Calibration protocol, attestation, certificates | ☑ Yes ❑ No |
| Operating manuals of the components | ☑ Yes ❑ No |
| EC-Declaration of conformity of components | ☑ Yes ❑ No |
| Operating manual:<br>☑ DE    original<br>❑ FR    translation<br>❑ EN    translation<br>❑ IT    translation<br>❑ ES    translation | ☑ Yes ❑ No |
| FAT,  factory acceptance test | ☑ Yes ❑ No |
| Acceptance protocol | ☑ Yes ❑ No |
| Handover certificate | ☑ Yes ❑ No |
| EC-Declaration of conformity | ☑ Yes ❑ No |

## 12.10 Content of the client's documentation

| Technical documentation | Required |
|---|---|
| Circuit layout | ☑ Yes ❏ No |
| Layout | ☑ Yes ❏ No |
| Layout chart | ☑ Yes ❏ No |
| R&I Schema | ❏ Yes ☑ No |
| Floor loading plan | ☑ Yes ❏ No |
| Pneumatics schema | ☑ Yes ❏ No |
| Hydraulics schema | ☑ Yes ❏ No |
| Detail drawing | ❏ Yes ☑ No |
| Calculations and experimental results | ☑ Yes ❏ No |
| Calibration protocol, attestation, certificates | ☑ Yes ❏ No |
| Operating manuals of the components | ☑ Yes ❏ No |
| EC-Declaration of conformity of the components | ☑ Yes ❏ No |
| Operating manual:<br>☑ DE     original<br>❏ FR     translation<br>❏ EN     translation<br>❏ IT     translation<br>❏ ES     translation | ☑ Yes ❏ No |
| FAT, factory acceptance test | ☑ Yes ❏ No |
| Acceptance protocol | ☑ Yes ❏ No |
| Handover certificate | ☑ Yes ❏ No |
| EC-Declaration of conformity | ☑ Yes ❏ No |

## 12.11 Release of the conformity assessment

| | |
|---|---|
| The Project manager | René Muster |
| Date | 10.03.2013 |
| Signature | |
| The CE-Delegate | Hans Muster |
| Date | 10.03.2013 |
| Signature | |

## 12.12 Acceptance protocol of a machine

| | |
|---|---|
| Description of the Machine | *Stanzmachine* |
| Type | *ST1* |
| Year of manufacture | *2013* |
| Serial number | *132654789EZB* |
| Client | *Stanztechnik AG, Stanzerstrasse 1, 123456 Zurich* |
| Manufacturer | *Mustermann AG Musterstr. 5 123456 Zurich* |
| Project manager | *René Muster* |
| CE Delegate | *Hans Muster* |
| In charge of technical documentation | *Hans Muster* |

**Client**

| Forename, Surname | Position | Signature |
|---|---|---|
| *Thomas Meier* | *Technics Manager* | _____ |
| *Theter Müller* | *Shop floor Manager* | _____ |
| *Joseph Fischer* | *Operator* | _____ |

**Stanztechnik AG**

| Forename, Surname | Position | Signature |
|---|---|---|
| *René Muster* | *Project Manager* | _____ |
| *Toni Polster* | *Service engineer* | _____ |
| *Thomas Walter* | *Service engineer* | _____ |

## General checkpoints

| Description | Fulfilled | Date Abbreviation |
|---|---|---|
| Machine delivered in full | ❑ Yes<br>❑ No | |
| Machine in good condition | ❑ Yes<br>❑ No | |
| Accessories available and complete | ❑ Yes<br>❑ No | |
| Accessories in good condition | ❑ Yes<br>❑ No | |
| Operating manual available (original and translation) | ❑ Yes<br>❑ No | |
| EC-Declaration of conformity available | ❑ Yes<br>❑ No | |
| Documentation specifically for this client is available | ❑ Yes<br>❑ No | |
| Measures L x W X H of machine according to layout chart | ❑ Yes<br>❑ No | |

## Function test

| Description | Fulfilled | Date Abbreviation |
|---|---|---|
| Machine installed | ❑ Yes<br>❑ No | |
| Required material connected | ❑ Yes<br>❑ No | |
| Test run according to test protocol | ❑ Yes<br>❑ No | |
| Client trained for safety devices | ❑ Yes<br>❑ No | |
| Emergency shutdown test performed | ❑ Yes<br>❑ No | |

| Description | Fulfilled | Date Abbreviation |
|---|---|---|
| Client was informed of the risks involved in the use of the Machine | ❏ Yes<br>❏ No | |
| Customer training performed and documented | ❏ Yes<br>❏ No | |
| Performance test | ❏ Yes<br>❏ No | |

## Deviations

Following deviations were ascertained and evaluated. The rectification of deviations is performed within the following terms by each person in charge

| Description of deviations | In charge / Deadline |
| --- | --- |
| | |
| | |
| | |
| | |
| | |
| | |
| | |

## Final evaluation

❑ All checkpoints treated fulfil the acceptance criteria, no deviation was ascertained

❑ Deviations are registered and must be sorted out prior to production start

❑ Critical failures: production of the Machine cannot be approved

Further remarks:

_____

_____

_____

_____

_____

## Client

| Forename, Surname | Position | Signature |
|---|---|---|
| Thomas Meier | Technics manager | |
| Theter Müller | Shop floor manager | |
| Joseph Fischer | Operator | |

## Stanztechnik AG

| Forename, Surname | Position | Signature |
|---|---|---|
| René Muster | Project Manager | |
| Toni Polster | Service Manager | |
| Thomas Walter | Service technician | |

# 13.    Risk assessment

The risk assessment serves as proof that a manufacturer respected all safety and health protective requirements.

Excerpt from the Machinery Directive 2006/42/EC:
- The manufacturer of machinery or his authorised representative must ensure that a risk assessment is carried out in order to determine the health and safety requirements which apply to the machinery. The machinery must then be designed and constructed taking into account the results of the risk assessment.

By the iterative process of risk assessment and risk reduction referred to above, the manufacturer or his authorised representative shall:
- determine the limits of the machinery, which include the intended use and any reasonably foreseeable misuse thereof,
- identify the hazards that can be generated by the machinery and the associated hazardous situations,
- estimate the risks, taking into account the severity of the possible injury or damage to health and the probability of its occurrence,
- evaluate the risks, with a view to determining whether risk reduction is required, in accordance with the objective of this Directive,
- eliminate the hazards or reduce the risks associated with these hazards by application of protective measures

The risk assessment must take into consideration tasks which need to be interrupted while taking protective measures, for example, when fixing or maintaining a machine.

Machinery should designed so that safety devices are provided, not only to the operator but also to any person in charge of installing, training, modifying or maintaining, without being obliged to interrupt their task.

These tasks must be determined by the risk assessment and considered as a part of the machine use.

## 13.1    FMEA

The FMEA „Failure Mode and Effects Analysis" is one of the most famous methods to ascertain and evaluate failures. It is adapted to evaluate machinery hazards.

Example of a hazard determination and failure evaluation according to ISO 12100: 2010

| Risk assessment | Stanzmachine ST1<br>Baujahr 2013<br>Serien Nr.132654789EZB | Mustermann AG<br>Musterstrasse 5<br>123456 Zurich |
|---|---|---|

| Probability of occurrence | Injury risk | Probability of detection | Risk priority number RPN |
|---|---|---|---|
| 1 very low<br>2 low<br>3 medium<br>4 high<br>5 very high | 1 none<br>2 low<br>3 high<br>4 disability<br>5 death | 1 High<br>2 moderate<br>3 low<br>4 very low<br>5 improbable | RPN <= 16 no action<br>RPN > 16 and <= 26<br>action required<br>RPN > 27 bis 125 action<br>to be defined |

| Hazard determination | | | | | | | | Taken action | | | | | |
|---|---|---|---|---|---|---|---|---|---|---|---|---|---|
| Hazards according to ISO 12100: 2010 | Danger zone | Source and possible consequences | Durability | occurrence | Injury risk | Detection | RPN | Constructive or technical protective measures | User information entry in operating manual | Occurrence | Injury risk | Detection | RPN |
| Electrical Hazards<br><br>Parts under voltage | Switch-board | Electric shock | 1<br>2<br>3 | 3 | 5 | 3 | 45 | Type of protec-tion IP2x | Device must be detached from power supply when working at switchboard | 1 | 2 | 3 | 6 |

**Approval and release of the risk assessment**

**Project manager**           René Muster
Date                              10.03.2013

Signature

**CE-Delegate**               Hans Muster
Date                              10.03.2013

Signature

Following divisions can be used to estimate the likelihood of hazard occurrence:
- very low
- low
- medium
- high
- very high

Following divisions can be used to attribute the injury risk and estimate the consequences of a hazard
- no risk of injury
- low  injury risk
- high injury risk
- disablement
- death

Following divisions can be used to estimate the probability of detecting a hazard:
- high
- medium
- low
- very low
- improbable

Procedure to determine hazards:
- Create your own template and include all hazards from ISO 12100
- Identify individual hazards together with the manufacturer
- Hazards which are not relevant or applicable are deleted
- Once the hazards have been identified, danger zones, their origin and consequences are to be listed
- Numbers 1, 2, 3, are indicated in the column "durability", (see example)
- The likelihood of occurrence, meaning, consequences, as well as the probability of detecting are now evaluated

- Numbers indicated on columns "occurrence", "injury risk" and "detection" are multiplied and the result registered on the RPN „Risk priority number" column.
  - RPN <= 16 no action required
  - RPN > 16 and <= 26 it has to be determined if any action is required
  - RPN > 27 to 125 action to be taken must be defined

Procedure to determine and evaluate measures
- Define constructive measures to eliminate hazards
- Technical and protective measures must be defined
- Residual risks are to be indicated in the column „user information" and must be included in the operating manual
- likelihood of occurrence, injury risk, consequences and probability of detecting can now be evaluated
- Numbers indicated on columns "occurrence", "injury risk" and "detection" are multiplied and the result registered on the RPN „Risk priority number"column.
  - RPN <= 16 no action required
  - RPN > 16 and <= 26 it has to be determined if any action is required
  - RPN > 27 to 125 action to be taken must be defined

Numbers 1, 2, and 3 are indicated in the column durability:
1. hazard during transportation, assembly or placing on the market
2. hazard during operation of the machine
3. hazard occurring between disassembly and dumping of the machine

## 13.2 Warning indicators against residual risks

If after these measures residual risks remain, the machine must have warning indicators which are visible, readable and permanently affixed.

Information and indicators on the machine should rather be presented in form of easily understandable symbols or pictograms.

Written and verbal information and indicators should be affixed and drafted in the official community language(s) where the machine is put on the market.

Warning indicators should be explained in the operating manual, for example:

 High voltage, electric voltage

 Radioactive material

 Caustic material

 Danger

 Danger of tripping

# 14. Design and construction of machinery

Machinery must be designed and constructed so that it is fitted for its function, and can be operated, adjusted and maintained without putting persons at risk when these operations are carried out under the conditions foreseen but also taking into account any reasonably foreseeable misuse thereof.

The aim of action taken must be to eliminate any risk throughout the foreseeable lifetime of the machinery including the phases of transport, assembly, dismantling, disabling and scrapping.

In selecting the most appropriate methods, the manufacturer or his authorised representative must apply the following principles, in the order given:
- eliminate or reduce risks as far as possible (inherently safe machinery design and construction),
- take the necessary protective measures in relation to risks that cannot be eliminated,
- inform users of the residual risks due to any shortcomings of the protective measures adopted, indicate whether any particular training is required and specify any need to provide personal protective equipment.

The machinery must be designed and constructed in such a way as to prevent abnormal use if such use would engender a risk. Where appropriate, the instructions must draw the user's attention to ways — which experience has shown might occur — in which the machinery should not be used.

Machinery must be designed and constructed to take account of the constraints to which the operator is subject as a result of the necessary or foreseeable use of personal protective equipment.

Machinery must be supplied with all the special equipment and accessories essential to enable it to be adjusted, maintained and used safely.

# 15.    Technical Documentation

Excerpt aus the Directive 768/2008/EC:

- The manufacturer shall establish the technical documentation. The documentation shall make it possible to assess the product's conformity to the relevant requirements, and shall include an adequate analysis and assessment of the risk(s). The technical documentation shall specify the applicable requirements and cover, as far as relevant for the assessment, the design, manufacture and operation of the product.

Each instruction manual must contain, where applicable, at least the following information:

- The business name and full address of the manufacturer and of his authorised representative;
- The designation of the machinery as marked on the machinery itself, except for the serial number (see section 1.7.3);
- The EC declaration of conformity, or a document setting out the contents of the EC declaration of conformity, showing the particulars of the machinery, not necessarily including the serial number and the signature;
- A general description of the machinery;
- The drawings, diagrams, descriptions and explanations necessary for the use, maintenance and repair of the machinery and for checking its correct functioning;
- A description of the workstation(s) likely to be occupied by operators;
- A description of the intended use of the machinery;
- Warnings concerning ways in which the machinery must not be used that experience has shown might occur;
- Assembly, installation and connection instructions, including drawings, diagrams and the means of attachment

and the designation of the chassis or installation on which the machinery is to be mounted;

- Instructions relating to installation and assembly for reducing noise or vibration;
- Instructions for the putting into service and use of the machinery and, if necessary, instructions for the training of operators;
- Information about the residual risks that remain despite the inherent safe design measures, safeguarding and complementary protective measures adopted;
- Instructions on the protective measures to be taken by the user, including, where appropriate, the personal protective equipment to be provided;
- The essential characteristics of tools which may be fitted to the machinery;
- The conditions in which the machinery meets the requirement of stability during use, transportation, assembly, dismantling when out of service, testing or foreseeable breakdowns;
- Instructions with a view to ensuring that transport, handling and storage operations can be made safely, giving the mass of the machinery and of its various parts where these are regularly to be transported separately;
- The operating method to be followed in the event of accident or breakdown; if a blockage is likely to occur, the operating method to be followed so as to enable the equipment to be safely unblocked;

The technical documentation should not be mixed up with the client's documentation. The client's documentation generally includes only the operating instructions and the EC-Declaration of conformity.

Technical documentation must not be in territory of the community or always physically available. However, it must be compiled and be made available within a reasonable deadline (according to its complexity) by the person in charge indicated on the EC-Declaration of conformity.

The match between a machine and the basic security and health requirements can be contested if the technical documentation is not made available to national examining authorities, further to a reasoned request.

## 15.1   Operating manual as per Directive 2006/42/EC

Operating manual

- Instructions must be drafted in one or more official Community languages. The words 'Original instructions' must appear on the language version(s) verified by the manufacturer or his authorised representative.

- In case of machinery intended to be used by non-professional operators, the wording and layout of the instructions must take into account the level of general education and acumen that can reasonably be expected from such operators

Understandability
- An operating manual including security information must accompany the product. It should be in a language that consumers or end users can easily understand. It is recommended to write in the national language of end users.

Translation
- The operating manual accompanying the machinery must be an "original operating manual" or a "translation of the original operating manual", in the second case, the original operating manual is to be enclosed.

Content of an operating manual according Directive 2006/42/EC
- business name and full address of the manufacturer and of his authorised representative
- designation of the machinery as marked on the machinery itself, except for the serial number
- EC declaration of conformity, or a document setting out the contents of the EC declaration of conformity, showing the particulars of the machinery but not necessarily including the serial number and the signature too
- general description of the machinery
- drawings, diagrams, descriptions and explanations necessary for the use, maintenance and repair of the machinery and for checking its correct functioning;
- a description of the probable workstation(s) of operators;
- a description of the intended use of the machinery;
- warnings -based on experience- concerning the ways in which the machine is not to be used;
- assembly, installation and connection instructions, including drawings, diagram, means of attachment and indications about the supporting framework or plant where the machinery is to be mounted;
- instructions relating installation and assembly in order to reduce noise or vibration; instructions for the putting into

service and for use of the machinery; if necessary, instructions for the training operators;

- information about residual risks that still remain, despite the included safe design measures, as well as safeguarding and complementary protective measures adopted;
- instructions are to be provided concerning protective measures to be taken by the user and, if necessary, about the personal protective equipment;
- essential characteristics of tools which may be fitted to the machinery;
- conditions in which the machinery achieves stability during use, transportation, assembly, dismantling, or when out of service, by testing or foreseeable breakdowns;
- instructions aiming safe transportation, handling and storage by giving the mass of the machinery and of its various parts if these are regularly to be transported apart;
- operating method to be followed in the event of accident or breakdown; if a blockage is likely to occur, the operating method to be followed in order to let the equipment be safely unblocked;

Further recommendations:
- a photo of the type plate
- warning and security indicators
- machinery layout, space requirement
- technical data in tabular form
- a description of the safety devices
- procedure by changes and building alterations
- extend of delivery
- training of enterprise
- PPE, personal protective equipment
- information about possible residual risks
- a copy of the EC-Declaration of conformity

DIN EN 62079 describes the content of an operating manual in detail. Its succeeding standard is the DIN EN 82079-1

## 15.2    Errors in the operating manual

Mistakes in the operating manual which risk safety and health of end user are considered as a product defect.

## 15.3    Drafting of the operating manual

Font type and size must ensure the best legibility; colours, symbols and a large display should be used to highlight indicators.

Operating manuals must be drafted in the language of the country where machinery is to be used for the first time and the original version should be indicated. If more than one language is required, each one of these should be distinctive. Illustrations should remain near to their corresponding text on translated versions.

Illustrations (photos) should complete the text to help comprehension. They should include an explanation, i.e. localization and identification of the manual control, appear next to the corresponding text and stick to the workflow.

For a better understanding, further information can be provided in tabular form. It should remain near the corresponding part.

The use of colours should be considered, particularly for devices needing a quick recognition.

If the operating manual is extensive, it should include a table of contents or an index.

The operation manual must be concise and easily understandable. Denominations and units should be have contents of the instructions must cover not only the intended use of the machinery but also take into account any reasonably foreseeable misuse thereof.

In the case of machinery intended for use by non-professional operators, the wording and layout of the instructions for use must take into account the level of general education and acumen that can reasonably be expected from such operators.

## 15.4   Attesting documentation

The technical documentation attests that the product placed on the market complies with all legal requirements.

## 15.5   Market observation

Manufacturers are responsible for monitoring their products, for analysing and taking corrective actions if any damage notification comes from a client, their own staff or any third person.

Within the scope of a QM system implementation as per ISO 9001, a complaint management is required, as a part of the standard.

## 15.6   Foodstuffs, cosmetic, pharmaceutical products

The instructions for foodstuffs machinery and machinery for use with cosmetics or pharmaceutical products must indicate recommended products and methods for cleaning, disinfecting and rinsing, not only for easily accessible areas but also for areas to which access is impossible or inadvisable.

# 16.    PPE: personal protective equipment

If during the risk assessment it becomes clear that personal protective equipment is necessary, labelling must be affixed to the machine, and/or be mentioned in the operating manual, and it should be included by trainings.

If known in advance that a special mark or a particular type of personal protective equipment is necessary, the client has to be informed prior to delivery in order to have enough time to acquire the PPE.

# 17. Content of the EC-Declaration of conformity

With the EC-Declaration of conformity a manufacturer declares that his product satisfies the corresponding standards and guidelines. The EC-Declaration of conformity is a statement at manufacturer's own responsibility.

There is no guideline for the creation of the document. The following information must be included in the EC-Declaration of conformity for machinery:

- declaration sentence: we, manufacturer, declare that the machinery described hereafter satisfies the following guidelines and norms;
- full address of the manufacturer;
- authorised representative for technical documentation;
- description of the machine, serial number, model, type;
- year of manufacture;
- used guidelines with date of issue;
- used standards with date of issue;
- name and function of the person signing the EC-Declaration of conformity, for example CE-Delegate, manager;
- signature of the CE-Delegate;
- place, Date

An EC-Declaration of conformity is issued for each machine and attached upon delivery. The EC-Declaration of conformity for integrated components, elements or subgroups is a part of the technical documentation.

The EC-Declaration of conformity helps authorities to verify if the corresponding guidelines and standards have been respected.

The EC-Declaration of conformity becomes increasingly accepted outside the EU.

The declaration must be issued in one of the official community languages in the EU (English, German, French, and Italian). Basically, the official language of the client should be chosen, for a better understanding.

## 17.1    EC-Declaration of conformity

We, manufacturer, declare that the machinery described hereafter satisfies the following guidelines and norms

| | |
|---|---|
| **Manufacturer** | Mustermann AG<br>Musterstrasse 5<br>123456 Zurich |
| **Authorised representative for technical documentation** | Hans Muster |
| **Description of the machine** | Punching machine |
| **Model, type** | ST1 |
| **Serial number** | 132654789EZB |
| **Year of manufacture** | 2013 |

| **Directives** | **Date of issue** |
|---|---|
| 2006/42/EC | 2006 |
| 94/9/EC | 1994 |

| **Standards** | **Date of issue** |
|---|---|
| EN 12100,    Risk assessment and Risk reduction | 2010 |
| EN 13849-1    safety-related parts of control systems | 2008 |
| EN 60204-1    electrical equipment of machines | 2006 |

..............................................    ........................

CE-Delegate                            Zurich
Hans Muster

# 18.    Retention period for documents

EC-Declaration of conformity ....................................................10 years
Assembly instructions...............................................................10 years
Operating manual......................................................................10 years
Conformity assessment.............................................................10 years
Risk assessment........................................................................10 years
Drawings, sketches, schemes ...................................................10 years
Calculations ..............................................................................10 years
Protocols....................................................................................10 years
Inspection reports ....................................................................10 years
EC type examination certificate...............................................15 years


If documents are stored in electronic form, they must remain available according to the guidelines.

When documents from a third company are stored, they must be covered by a contract.

# 19. Terms of the Directive 2006/42/EC

Placing on the market
- placing on the market' means making available for the first time in the Community machinery or partly completed machinery with a view to distribution or use, whether for reward or free of charge;

Manufacturer
- manufacturer' means any natural or legal person who designs and/or manufactures machinery or partly completed machinery covered by this Directive and is responsible for the conformity of the machinery or the partly completed machinery with this Directive with a view to its being placed on the market, under his own name or trademark or for his own use. In the absence of a manufacturer as defined above, any natural or legal person who places on the market or puts into service machinery or partly completed machinery covered by this Directive shall be considered a manufacturer;

Authorised representative
- authorised representative' means any natural or legal person established in the Community who has received a written mandate from the manufacturer to perform on his behalf all or part of the obligations and formalities connected with this Directive;

Putting into service
- putting into service' means the first use, for its intended purpose, in the Community, of machinery covered by this Directive;

## Harmonised Standard

- harmonised standard' means a non-binding technical specification adopted by a standardisation body, namely the European Committee for Standardisation (CEN), the European Committee for Electrotechnical Standardisation (CENELEC) or the European Telecommunications Standards Institute (ETSI), on the basis of a remit issued by the Commission in accordance with the procedures laid down in Directive 98/34/EC of the European Parliament and of the Council of 22 June 1998 laying down a procedure for the provision of information in the field of technical standards and regulations and of rules on Information Society services

## Hazard

- hazard' means a potential source of injury or damage to health

## Danger zone

- 'danger zone' means any zone within and/or around machinery in which a person is subject to a risk to his health or safety

## Exposed person

- 'exposed person' means any person wholly or partially in a danger zone

## Operator

- operator' means the person or persons installing, operating, adjusting, maintenance, cleaning, repairing or moving machinery.

## Risk

- risk' means a combination of the probability and the degree of an injury or damage to health that can arise in a hazardous situation;

Guard
- guard' means a part of the machinery used specifically to provide protection by means of a physical barrier

Protective device
- 'protective device' means a device (other than a guard) which reduces the risk, either alone or in conjunction with a guard

Intended use
- intended use' means the use of machinery in accordance with the information provided in the instructions for use;

Reasonably foreseeable misuse
- reasonably foreseeable misuse' means the use of machinery in a way not intended in the instructions for use, but which may result from readily predictable human behaviour.

# 20.    Summary

### Risk assessment

- Determine risks
- Evaluate risks
- Determine actions
- Evaluate actions

### Conformity assessment

- Guidelines
- Standards
- Limits of the Machine
- Stages of life

### Technical Documentation

- Drawings
- Schemes, Sketches
- Operating manual
- Calculations
- Technical reports
- Certificates
- Acceptance protocols

### EC-Declaration of conformity

### CE marking   affix

### Minimum client  documentation

- EC-Declaration of conformity
- Operating manual

# 21. Quality management as per ISO 9001: 2008

## 21.1 Certification procedure

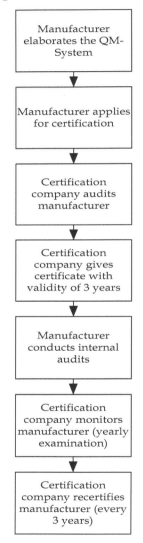

In order to implement a complete Quality management according to ISO 9001: 2008, the following must be accomplished:

## 21.2   Quality Management System

QM-Requirements

- necessary processes and their application in the organization are determined,
- sequence and interaction of these processes is established,
- necessary criteria and applicable methods are defined to ensure that both the operation and control of these processes are effective,
- resources to support the processes are available,
- processes are monitored, analysed and evaluated,
- actions to achieve planned results are decided,
- outsourced processes are directed and monitored

Documentation requirements

- documented statements of a quality policy and quality objectives,
- a quality manual,
- documented procedures and records required are available,

Control of documents

- documents are approved,
- present document status is clearly indicated,
- relevant versions of applicable documents are available at points of use,
- documents remain legible and readily identifiable,
- documents of external origin determined by the organization to be necessary for the planning and operation of the

quality management system are identified and their distribution controlled, and

- prevent the unintended use of obsolete documents, and to apply suitable identification to them if they are retained for any purpose.

Control of records

The organization shall establish a documented procedure to define the controls needed for the identification, storage, protection, retrieval, retention and disposition of records.

Records shall remain legible, readily identifiable and retrievable

## 21.3    Responsibility of the Management

The organization shall determine and provide the resources needed

- to implement and maintain the quality management system and continually improve its effectiveness, and
- to enhance customer satisfaction by meeting customer requirements.

Customer focus

- customer's needs are noted, understood and implemented.

Quality policy

- the quality policy is communicated and understood.

Planning

- quality objectives to fulfil the requirements for product are established.
- the planning of the quality management system is carried out in order to meet requirements and quality objectives

Authority and communication

- authority and communication are ensured and known.
- a management member is designated as responsible of the system

- communication processes are available and communication takes place regarding the effectiveness of the quality management system

Management review
  Management review is carried out and documented regularly

  A management review includes the following information:
- results of the audits,
- customer feedback,
- process performance and product conformity,
- status of preventive and corrective actions,
- follow-up actions from previous management reviews,
- changes that could affect the quality management system and
- recommendations for improvement

## 21.4    Resource management

Provision of resources
  The organization shall determine and provide the resources needed to maintain the level of QMS and to meet client's requirements.

Human resources

  Personnel shall be competent on the basis of appropriate education, training, skills and experience.

- necessary skills for personnel are determined
- training is provided to achieve necessary competence,
- personnel has the adequate education (certificate available),
- effectiveness of training is evaluated,

- personnel are aware of the relevance and importance of their activities and how they contribute to the achievement of the quality objectives,
- Appropriate records of education, training, stills and experience.

Infrastructure

The organisation shall determine, provide and maintain the infrastructure needed to achieve conformity to product requirements (buildings, workspace, hardware, software, supporting services such as transport, communication, etc.).

Work environment

The organization shall determine and manage the work environment needed to achieve conformity to product requirements.

## 21.5 Product realization

Planning and product realization

Processes are determined or developed, as appropriate.
- quality objectives and requirements for the product.
- the need to establish processes and documents, and to provide resources which are specific to the product;
- required verification, validation, monitoring, measurement, inspection and test activities specific to the product and the criteria for product acceptance;
- records needed to provide evidence that the realization processes and resulting product meet requirements.

## 21.6    Customer-related process

- requirements of the client are specified,
- statutory and regulatory requirements are determined,
- feasibility of the client's requirements is evaluated and   registered,
- differences between statutory and regulatory requirements and those specified by the customer are established.

Arrangements for communication with customers are determined concerning

- product information
- enquiries, contracts or order handling, including amendments, and
- customer feedback, including customer complaints.

## 21.7    Development

The organization shall plan and control the development of product. It shall determine

  a) development stages,
  b) review, verification and validation that are appropriate to each design and development stage, and

- interfaces between different groups involved and responsibility clearly assigned,
- development target includes all requirements,
- design and development review are systematically carried out.,
- design and development outputs are verifiable,
- development is validated according to provable criteria,
- changes are clearly documented and distinguishable. They include any necessary reenactment for further development.

## 21.8    Purchasing

Deliveries suffice requirements of end product. Products and deliveries are chosen or evaluated according to defined criteria.

Purchasing corresponds to clear specifications. Purchased products are tested according to defined criteria and approved for use.

## 21.9    Manufacture and Services

- manufacture takes place under controlled conditions,
- manufacture procedures are clearly defined and validated or checkable,
- products are allocated, marked and traceable if necessary
- property of client is declared and protected,
- conformity of product is preserved during internal processing and delivery

## 21.10  Control of monitoring and measuring equipment

Necessary monitoring and measuring equipment are determined.

Procedures to control monitoring and measuring equipment are available.

Measuring equipment ought to be:
- periodically or prior to measurement calibrated,
- if necessary adjusted or readjusted
- marked,
- safeguarded from adjustments that would invalidate the measurement result,
- protected from damage

Results of measuring are recorded and evaluated. Calibration and verification are documented.

## 21.11   Measurement, Analysis and Improvement

Miscellaneous

Procedures to monitor, measure, analyse and improve are planned and implemented.

Surveillance and measurement entails:
- a defined procedure to determine and implement client's opinion,
- a defined procedure to plan and carry out internal audits in order to monitor the QM-System performance, and to keep tracking of findings,
- surveillance and measurement of manufacturing procedures as well as implementation of corrective actions,
- surveillance and measurement of product characteristics, proof of acceptance

## 21.12   Control of nonconforming products

The organization shall ensure that product which does not conform to product requirements is identified and controlled to prevent its unintended use or delivery. Procedures must be defined
- to eliminate a detected nonconformity,
- to use, release or accept after a special release,
- to destroy nonconforming products,

Records are kept concerning type of nonconformity and take actions.

Corrective actions are taken and documented concerning nonconforming products

## 21.13  Analysis of data

Data concerning customer satisfaction, product quality, suppliers, QM-System are properly gathered and analysed.

## 21.14  Improvement

Impact of the system should be constantly and measurably improved. This means:

A procedure must be defined to document, analyse and evaluate detected nonconformities and performed corrective actions.

A procedure must be defined to prevent nonconformities. Its implementation and extend are to be documented and evaluated.

# 22.  Second-hand machinery

Second-hand machinery imported from a third country and second-hand machinery arriving into the Community market for the first time must conform to machinery guidelines.

This is <u>not</u> applicable for machinery already placed on the market within the Community.

The obligation to affix the CE marking extends to all machines subject to the machinery directive.

This is valid for:
- new machinery, irrespective of manufactured within the Community or in a third country;
- second-hand machinery imported from a third country;
- second-hand machinery in general;
- machines with major modifications

If a modified machine is considered to be a new product, the related guidelines must be observed when placing on the market and putting into operation.

This shall be verified with the corresponding conformity assessment, indicated in the related guideline. This process will be carried out only if considered necessary after a risk assessment.

## 23.    Second-hand machinery classified as new

Excerpt of the Blue Guide:

if the risk assessment leads to the conclusion that the nature of the hazard or the level of risk has increased, then the modified product should normally be considered as a new product.

The person who carries out important changes to the product is responsible for verifying whether or not it should be considered as a new product.

## 24.    Building alterations, self-building

Machines which have been subject to changes or being self-built must fulfill machinery guidelines.

For modified or self-built machines, a risk assessment should be carried out in order to insure health protection

Excerpt from dem Blue Guide:

A product, which has been subject to important  changes that aim to modify its original performance, purpose or type after it has been put into service, may be considered as a new product.

This has to be assessed on a case-by-case basis and, in particular, in view of the objective of the directive and the type of products covered by the directive in question.

# 25. Procedure for partly completed machinery

The Manufacturer of partly completed machinery or his authorised representative has to ensure what follows before placing the machine on the market:

- an appropriate technical documentation;
- assembly instructions;
- a declaration of incorporation;
- assembly instructions and declaration of incorporation enclosed to the partly completed machinery until full incorporation and finally included with the completed machine as a part of the technical documentation.

# 26.    Relevant technical Documentation

This part describes the procedure for compiling relevant technical documentation. These documents must clearly show which requirements of the directive are applicable and if they are being fulfilled.

The technical file must also cover design, manufacture and operation of the partly completed machinery, if this is considered necessary for assessing the compliance of the health and safety requirements that have been applied.

The documentation must be compiled in one or more official Community languages. It shall comprise the following:

- the overall drawing of the partly completed machinery and drawings of the control circuits,
- full detailed drawings, accompanied by any calculation notes, test results, certificates, etc., required to check the conformity of the partly completed machinery with the essential health and safety requirements that have been applied,
- risk assessment documentation showing the procedure followed, including:
    - a list of the essential health and safety requirements that have been applied and fulfilled,
    - the description of the protective measures implemented to eliminate identified hazards or to reduce risks and, where appropriate, the indication of the residual risks,
    - the standards and other technical specifications used, indicating the essential health and safety requirements covered by these standards,
    - any technical report giving the results of the tests carried out either by the manufacturer or

by a body chosen by the manufacturer or his authorised representative,
- a copy of the assembly instructions for the partly completed machinery;
- in case of series manufacture, the internal measures that shall be implemented to ensure that the partly completed machinery remains in conformity with the essential health and safety requirements that have been applied.

The manufacturer must carry out any necessary research and test on components and fittings of the partly completed machinery to determine if, given its design or construction, this one can be assembled and used safely. The relevant reports and results shall be included in the technical file.

All relevant technical documentation must remain readily available for at least 10 years after the date of manufacture of the partly completed machine, of a series manufacture, or a last unit production, and be provided to the competent authorities of the Member States, on request. Documents must not be located in the territory of the Community or be permanently available in physical form. However, they shall be assembled and presented to the relevant authority by the person designated in the declaration for incorporation.

Failure to present the relevant technical documentation in response to a duly reasoned request by the competent national authorities may constitute sufficient grounds for doubting that the health and safety requirements being applied and attested are conform.

## 26.1 Assembly instructions

The assembly instructions for partly completed machinery must contain a description of the conditions which must be met with a view to correct incorporation in the final machinery, so as not to compromise safety and health.

The assembly instructions must be written in an official Community language acceptable to the manufacturer of the machinery in which the partly completed machinery shall be assembled, or to his authorised representative.

## 26.2    Declaration of incorporation

This declaration and translations thereof must be drawn up under the same conditions as the instructions (see Annex 1, section 1.7.4.1(a) and (b)), and must be typewritten or else handwritten in capital letters.

The declaration of incorporation must contain the following particulars:

- business name and full address of the manufacturer of the partly completed machinery and, where appropriate, his authorised representative;
- name and address of the person authorised to compile the relevant technical documentation, who must be established in the Community;
- description and identification of the partly completed machinery including generic denomination, function, model, type, serial number and commercial name;
- a sentence declaring which essential requirements of this Directive are applied and fulfilled and that the relevant technical documentation is compiled in accordance with part B of Annex VII, and, where appropriate, a sentence declaring the conformity of the partly completed machinery with other relevant Directives. These references must be those of the texts published in the Official Journal of the European Union;
- an undertaking to transmit, in response to a reasoned request by the national authorities, relevant information on the partly completed machinery. This shall include the method of transmission and shall be without prejudice to

the intellectual property rights of the manufacturer of the partly completed machinery;

- a statement that the partly completed machinery must not be put into service until the final machinery into which it is to be incorporated has been declared in conformity with the provisions of this Directive, where appropriate;
- the place and date of the declaration;
- the identity and signature of the person empowered to draw up the declaration on behalf of the manufacturer or his authorised representative.

## 26.3 Example of an EC-Declaration of incorporation

As manufacturer of a partly completed machine, we declare that:
- the partly completed machine described hereafter fulfills the cited guidelines and standards
- adapted technical documentation has been issued
- the technical file shall be provided to authorities in paper or electronic form, on request

| | |
|---|---|
| **Manufacturer** | Mustermann AG<br>123456 Musterbach |
| **Authorised representative for technical documentation** | Hans Muster |
| **Description of the partly completed Machine** | Punching module |
| **Model, type** | SM1 |
| **Serial number** | 32145687 |
| **Year of manufacture** | 2013 |

| **Directive** | **Date of issue** |
|---|---|
| 2006/42/EC | 2006 |

| **Standards** | **Date of issue** |
|---|---|
| EN 12100, Risk assessment and risk reduction | 2010 |
| EN 13849-1, Safety-related parts of control systems | 2008 |

The partly completed machine can only be put into service after it has been proofed (if applicable), that the requirements of the guideline (Directive 2006/42/EC) have been respected for the machine which is to be integrated into the partly completed machine.

................................................... .................

CE-Delegate                                                    Zurich
Hans Muster

# 27.    The EC type examination

The type examination is the procedure whereby a notified body ascertains and certifies that a representative model of machinery referred to in Annex IV (hereafter named the type) satisfies the provisions of this directive.

The manufacturer or his authorised representative must, for each type, draw up the technical file referred to in Annex VII, part A. For each type, the application for an EC type examination shall be submitted by the manufacturer or his authorised representative to a notified body of his choice.

```
┌─────────────────────────┐
│   Manufacturer applies   │
│     for an EC type-      │
│       examination        │
└─────────────────────────┘
             │
             ▼
┌─────────────────────────┐
│      Notified body       │
│ controls technical file  │
│       and type-          │
│       examination        │
└─────────────────────────┘
             │
             ▼
┌─────────────────────────┐
│   Notified body issues   │
│  EC type-examination     │
│       certificate        │
└─────────────────────────┘
             │
             ▼
┌─────────────────────────┐
│  Notified body makes     │
│  sure that EC type-      │
│ examination remains      │
│         valid            │
└─────────────────────────┘
             │
             ▼
┌─────────────────────────┐
│     Erstellen der        │
│         EG-              │
│  Konformitätserkläru     │
│          ng              │
└─────────────────────────┘
             │
             ▼
┌─────────────────────────┐
│  Manufacturer applies    │
│    for EC type-          │
│     examination          │
│ certificate every 5      │
│         years            │
└─────────────────────────┘
```

## 27.1   Application for an EC type examination

The application must include the following:
- Name and address of the manufacturer and, where appropriate, his authorized representative
- a written declaration that the application has not been submitted to another notified body
- the technical file

Moreover, the applicant shall place at the disposal of the notified body a sample of the type. The notified body may ask for further samples if the test programme so requires.

## 27.2   Notified body

The notified body shall:

- examine the technical file, check that the type was manufactured in accordance with it and establish which elements have been designed in accordance with the relevant provisions of the standards referred to in Article 7(2), and those elements whose design is not based on the relevant provisions of those standards;
- carry out or have carried out appropriate inspections, measurements and tests to ascertain whether the solutions adopted satisfy the essential health and safety requirements of this Directive, where the standards referred to in Article 7(2) were not applied;
- where harmonised standards referred to in Article 7(2) were used, carry out or have carried out appropriate inspections, measurements and tests to verify that those standards were actually applied;

- agree with the applicant as to the place where the check that the type was manufactured in accordance with the examined technical file and the necessary inspections, measurements and tests will be carried out

## 27.3    EC type examination certificate

If the type satisfies the provisions of this Directive, the notified body shall issue the applicant with an EC type examination certificate. The certificate shall include the name and address of the manufacturer and his authorised representative, the data necessary for identifying the approved type, the conclusions of the examination and the conditions to which its issue may be subject. The manufacturer and the notified body shall retain a copy of this certificate, the technical file and all relevant documents for a period of 15 years from the date of issue of the certificate. 5. If the type does not satisfy the provisions of this Directive, the notified body shall refuse to issue the applicant with an EC type-examination certificate, giving detailed reasons for its refusal. It shall inform the applicant, the other notified bodies and the Member State which notified it. An appeal procedure must be available.

The applicant shall inform the notified body which retains the technical file relating to the EC type examination certificate of all modifications to the approved type.

The notified body shall examine these modifications and shall then either confirm the validity of the existing EC type examination certificate or issue a new one if the modifications are liable to compromise conformity with the essential health and safety requirements or the intended working conditions of the type.

The Commission, the Member States and the other notified bodies may, on request, obtain a copy of the EC type examination certificates. On reasoned request, the Commission and the Member States may obtain a copy of the technical file and the results of the examinations carried out by the notified body.

## 27.4    Modifications to the approved type

The applicant shall inform the notified body which retains the technical file relating to the EC type examination certificate of all modifications to the approved type.

The notified body shall examine these modifications and shall then either confirm the validity of the existing EC type examination certificate or issue a new one if the modifications are liable to compromise conformity with the essential health and safety requirements or the intended working conditions of the type.

The Commission, the Member States and the other notified bodies may, on request, obtain a copy of the EC type examination certificates. On reasoned request, the Commission and the Member States may obtain a copy of the technical file and the results of the examinations carried out by the notified body

## 27.5 Official Community language of procedure

Files and correspondence referring to the EC type-examination procedures shall be written in the official Community language(s) of the Member State where the notified body is established or in any other official Community language acceptable to the notified body.

## 27.6 Validity of the Certificate

The notified body has the ongoing responsibility of ensuring that the EC type examination certificate remains valid. It shall inform the manufacturer of any major changes which would have an implication on the validity of the certificate. The notified body shall withdraw certificates which are no longer valid.

The manufacturer of the machinery concerned has the ongoing responsibility of ensuring that the said machinery meets the corresponding state of the art.

## 27.7 Review of the validity

The manufacturer shall request from the notified body the review of the validity of the EC type examination certificate every five years.

If the notified body finds that the certificate remains valid, taking into account the state of the art, it shall renew the certificate for a further five years.

The manufacturer and the notified body shall retain a copy of this certificate, of the technical file and of all the relevant documents for a period of 15 years from the date of issue of the certificate.

In the event that the validity of the EC type examination certificate is not renewed, the manufacturer shall cease the placing on the market of the machinery concerned.

## 28.    Product Safety Act

The Product Safety Act stipulates that a product shall not endanger safety and health of individuals. This applies for second-hand products as well.

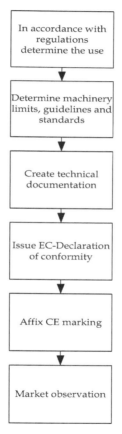

The Product Safety Act concerns all products placed on the market. Exceptions are: medical devices, comestible, antiques, chemicals, weapons, explosives, for which an adapted legislation is applicable.

# 29. Residual risks

Indicators should warn users against residual risks. Warnings must also be included and depicted in the operating manual.

# 30. Callback, Market observation

Even after placing on the market, complaints, reports or accidents can reveal safety problems. In such cases, the distributor must work out and implement the corresponding procedures.

## 30.1 Callback reporting obligation

A distributor shall report to authorities any dangerous product and any callback action.

# 31.    The "GS" tested safety mark

For the display of the GS-Mark, a dark font on a pale basis or a bright font on a dark basis are authorised. If the magnitude is changed, proportions are to be kept.

The Id-Mark ("Id-Zeichen"): GS-Mark and GS body are to be combined. The icon of the GS body should appear on the upper left hand side of the GS-Mark. If required, it can slightly extend beyond the margin of the GS-Mark to keep the general view of this one.

If the GS-Mark is 2 cm high or less, it is allowed to place the GS body left from the GS-Mark. In this case, however, the GS body icon must touch the GS-Mark in order to preserve the unit of the safety mark. The GS body cannot be bigger than the GS-Mark.

The GS-Mark can only be affixed after a type-examination. All required standards and guidelines must be observed and documented by the GS body.

The GS body shall prepare the certificate concerning the conformity of the GS-Mark. The conformity shall be valid for a maximum of five years and limited to a manufacture contingent. The GS body shall publish a list of the certificates issued and can suspend the conformity in case of reasonable doubts about the compliance of the GS-Mark.

# 32.    Market surveillance

Member States shall take all appropriate measures to ensure that machinery may be placed on the market and/or put into service only if it satisfies the relevant provisions of this Directive and does not endanger the health and safety of persons and, where appropriate, domestic animals or property, when properly installed and maintained and used for its intended purpose or under conditions which can reasonably be foreseen.

Member States shall take all appropriate measures to ensure that partly completed machinery can be placed on the market only if it satisfies the relevant provisions of this Directive.

Member States shall institute or appoint the competent authorities to monitor the conformity of machinery and partly completed machinery with the provisions set out in paragraphs 1 and 2.

Member States shall define the tasks; organisation and powers of the competent authorities referred to in paragraph 3 and shall notify the Commission and other Member.

# 33. Notified bodies

Member States shall notify the Commission and the other Member States of the bodies which they have appointed to carry out the assessment of conformity for placing on the market referred to in Article 12(3) and (4), together with the specific conformity assessment procedures and categories of machinery for which these bodies have been appointed and the identification numbers assigned to them beforehand by the Commission. Member States shall notify the Commission and other Member States of any subsequent amendment.

The Member States shall ensure that the notified bodies are monitored regularly to check that they comply at all times with the criteria set out in Annex XI. The notified body shall provide all relevant information on request, including budgetary documents, to enable the Member States to ensure that the requirements of Annex XI are met.

Member States shall apply the criteria set out in Annex XI in assessing the bodies to be notified and the bodies already notified.

The Commission shall publish in the Official Journal of the European Union, for information, a list of the notified bodies and their identification numbers and the tasks for which they have been notified. The Commission shall ensure that this list is kept up to date.

Bodies meeting the assessment criteria laid down in the relevant harmonised standards, the references of which shall be published in the Official Journal of the European Union, shall be presumed to fulfil the relevant criteria.

If a notified body finds that relevant requirements of this Directive have not been met or are no longer met by the manufacturer or that an EC type-examination certificate or the approval of a quality assurance system should not have been issued, it shall,

taking account of the principle of proportionality, suspend or withdraw the certificate or the approval issued or place restrictions on it, giving detailed reasons, unless compliance with such requirements is ensured by the implementation of appropriate corrective actions by the manufacturer.

In the event of suspension or withdrawal of the certificate or the approval or of any restriction placed on it, or in cases where intervention by the competent authority may prove necessary, the notified body shall inform the competent authority pursuant to Article 4. The Member State shall inform the other Member States and the Commission without delay. An appeal procedure shall be available. The Commission shall provide for the organisation of an exchange of experience between the authorities responsible for appointment, notification and monitoring of notified bodies in the Member States, and the notified bodies, in order to coordinate the uniform application of this Directive.


A Member State which has notified a body shall immediately withdraw its notification if it finds: (a) that the body no longer meets the criteria set out in Annex XI; or (b) that the body seriously fails to fulfil its responsibilities. The Member State shall immediately inform the Commission and the other Member States accordingly.

# 34. Health & safety requirements

The basic safety and health requirements are described in the Machinery Directive Annex I, here are some examples:

## 34.1 Control systems

Control systems must be designed and constructed in such a way as to prevent hazardous situations from arising. Above all, they must be designed and constructed in such a way that:

- they can withstand the intended operating stresses and external influences,
- a fault in the hardware or the software of the control system does not lead to hazardous situations,
- errors in the control system logic do not lead to hazardous situations,
- reasonably foreseeable human error during operation does not lead to hazardous situations.

Particular attention must be given to the following points:
- the machinery must not start unexpectedly,
- the parameters of the machinery must not change in an uncontrolled way, where such change may lead to hazardous situations,
- the machinery must not be prevented from stopping if the stop command has already been given,
- no moving part of the machinery or piece held by the machinery must fall or be ejected,
- automatic or manual stopping of the moving parts, whatever they may be, must be unimpeded,
- the protective devices must remain fully effective or give a stop command,

- the safety-related parts of the control system must apply in a coherent way to the whole of an assembly of machinery and/or partly completed machinery.

For cable-less control, an automatic stop must be activated when correct control signals are not received, including loss of communication.

## 34.2    Materials and products

The materials used to construct machinery or products used or created during its use must not endanger persons' safety or health. In particular, where fluids are used, machinery must be designed and constructed to prevent risks due to filling, use, recovery or draining.

## 34.3    Lighting

Machinery must be supplied with integral lighting suitable for the operations concerned where the absence thereof is likely to cause a risk despite ambient lighting of normal intensity.

Machinery must be designed and constructed so that there is no area of shadow likely to cause nuisance, that there is no irritating dazzle and that there are no dangerous stroboscopic effects on moving parts due to the lighting.

Internal parts requiring frequent inspection and adjustment, and maintenance areas must be provided with appropriate lighting.

.

## 34.4    Starting

It must be possible to start machinery only by voluntary actuation of a control device provided for the purpose.

The same requirement applies:

- when restarting the machinery after a stoppage, whatever the cause;
- when effecting a significant change in the operating conditions

## 34.5    Normal stop

Machinery must be fitted with a control device whereby the machinery can be brought safely to a complete stop.

Each workstation must be fitted with a control device to stop some or all of the functions of the machinery, depending on the existing hazards, so that the machinery is rendered safe.

The machinery's stop control must have priority over the start controls.

Once the machinery or its hazardous functions have stopped, the energy supply to the actuators concerned must be cut off.

## 34.6    Emergency stop

Machinery must be fitted with one or more emergency stop devices to enable actual or impending danger to be averted.

The following exceptions apply:
- machinery in which an emergency stop device would not lessen the risk, either because it would not reduce the stopping time or because it would not enable the special measures required to deal with the risk to be taken;
- portable hand-held and/or hand-guided machinery.

The device must:
- have clearly identifiable, clearly visible and quickly accessible control devices,
- stop the hazardous process as quickly as possible, without creating additional risks,

- where necessary, trigger or permit the triggering of certain safeguard movements.
- Once active operation of the emergency stop device has ceased following a stop command, that command must be sustained by engagement of the emergency stop device until that engagement is specifically overridden; it must not be possible to engage the device without triggering a stop command; it must be possible to disengage the device only by an appropriate operation, and disengaging the device must not restart the machinery but only permit restarting.
- The emergency stop function must be available and operational at all times, regardless of the operating mode.
- Emergency stop devices must be a back-up to other safeguarding measures and not a substitute for them

## 34.7   Lightning

Machinery in need of protection against the effects of lightning while being used must be fitted with a system for conducting the resultant electrical charge to earth.

## 34.8   Indicators and Alarm

In order to prevent from imminent hazards like unexpected start or overspeed, the use of indicators like a blinking light and an acoustic alarm is allowed.

These warning signals may be also used to prevent users before any protective measure is activated.

It is important that these warnings:
- come prior to danger,
- are clear,
- are recognizable and can be distinguished from other warning signals,
- are easily recognizable by user or other persons

Warning devices must be designed and placed so that they can be easily checked.

User information must prescribe the regular verification of warning devices.

Constructors must consider the possibility of an overstimulation caused by many visual and acoustic signals which could lead to overlooking warning signals.

## 34.9    Failure by the power supply

The interruption, the re-establishment after an interruption or the fluctuation in whatever manner of the power supply to the machinery must not lead to dangerous situations.

Particular attention must be given to the following points:
- The interruption, the re-establishment after an interruption or the fluctuation in whatever manner of the power supply to the machinery must not lead to dangerous situations
- The machinery must not start unexpectedly, the parameters of the machinery must not change in an uncontrolled way when such change can lead to hazardous situations,
- the machinery must not be prevented from stopping if the command has already been given,
- no moving part of the machinery or piece held by the machinery must fall or be ejected, automatic or manual stopping of the moving parts, whatever they may be, must be unimpeded, the protective devices must remain fully effective or give a stop command

## 34.10    Risks due to falling or ejected objects

Precautions must be taken to prevent risks from falling or ejected objects.

### 34.11    Risks due to surfaces, edges or angles

Insofar as their purpose allows, accessible parts of the machinery must have no sharp edges, no sharp angles and no rough surfaces likely to cause injury.

### 34.12    Risks related to moving parts

The moving parts of machinery must be designed and constructed in such a way as to prevent risks of contact which could lead to accidents or must, where risks persist, be fitted with guards or protective devices.

All necessary steps must be taken to prevent accidental blockage of moving parts involved in the work. In cases where, despite the precautions taken, a blockage is likely to occur, the necessary specific protective devices and tools must, when appropriate, be provided to enable the equipment to be safely unblocked.

The instructions and, where possible, a sign on the machinery shall identify these specific protective devices and how they are to be used.

### 34.13    Protection against risks arising from moving parts

Guards or protective devices designed to protect against risks arising from moving parts must be selected on the basis of the type of risk. The following guidelines must be used to help to make the choice

# 35. Requirements for safety devices

## 35.1 General requirements

Guards and protective devices must:

- be of robust construction,
- be securely held in place,
- not give rise to any additional hazard,
- not be easy to by-pass or render non-operational,
- be located at an adequate distance from the danger zone,
- cause minimum obstruction to the view of the production process, and
- enable essential work to be carried out on the installation and/or replacement of tools and for maintenance purposes by restricting access exclusively to the area where the work has to be done, if possible without the guard having to be removed or the protective device having to be disabled.

In addition, guards must, where possible, protect against the ejection or falling of materials or objects and against emissions generated by the machinery.

# 36. Risks arising from other hazards

## 36.1 Electricity supply

Where machinery has an electricity supply, it must be designed, constructed and equipped in such a way that all hazards of an electrical nature are or can be prevented.

The safety objectives set out in Directive 73/23/EEC shall apply to machinery. However, the obligations concerning conformity assessment and the placing on the market and/or putting into service of machinery with regard to electrical hazards are governed solely by this Directive.

## 36.2 Static electricity

Machinery must be designed and constructed to prevent or limit the build-up of potentially dangerous electrostatic charges and/or be fitted with a discharging system.

## 36.3 Energy supply other than electricity

Where machinery is powered by source of energy other than electricity, it must be so designed, constructed and equipped as to avoid all potential risks associated with such sources of energy.

## 36.4 Errors when fitting

Errors likely to be made when fitting or refitting certain parts which could be a source of risk, must be made impossible by the design and construction of such parts or, failing this, by information given on the parts themselves and/or their housings. The same information must be given on moving parts and/or their housings where the direction of movement needs to be known in order to avoid a risk.

Where necessary, the instructions must give further information on these risks.

Where a faulty connection can be the source of risk, incorrect connections must be made impossible by design or, failing this, by information given on the elements to be connected and, where appropriate, on the means of connection.

## 36.5  Extreme Temperature

Steps must be taken to eliminate any risk of injury arising from contact with or proximity to machinery parts or materials at high or very low temperatures.

The necessary steps must also be taken to avoid or protect against the risk of hot or very cold material being ejected.

## 36.6  Fire

Machinery must be designed and constructed in such a way as to avoid any risk of fire or overheating posed by the machinery itself or by gases, liquids, dust, vapours or other substances produced or used by the machinery.

## 36.7  Explosion

Machinery must be designed and constructed in such a way as to avoid any risk of explosion posed by the machinery itself or by gases, liquids, dust, vapours or other substances produced or used by the machinery.

Machinery must comply, as far as the risk of explosion due to its use in a potentially explosive atmosphere is concerned, with the provisions of the specific Community Directives.

## 36.8  Noise

Machinery must be designed and constructed in such a way that risks resulting from the emission of airborne noise are reduced to the lowest level, taking account of technical progress and the availability of means of reducing noise, in particular at source.

The level of noise emission may be assessed with reference to comparative emission data for similar machinery.

### 36.9    Vibrations

Machinery must be designed and constructed in such a way that risks resulting from vibrations produced by the machinery are reduced to the lowest level, taking account of technical progress and the availability of means of reducing vibration, in particular at source.

The level of vibration emission may be assessed with reference to comparative emission data for similar machinery.

### 36.10    Radiation

Undesirable radiation emissions from the machinery must be eliminated or be reduced to levels that do not have adverse effects on persons.

Any functional ionising radiation emissions must be limited to the lowest level which is sufficient for the proper functioning of the machinery during setting, operation and cleaning. Where a risk exists, the necessary protective measures must be taken.

Any functional non-ionising radiation emissions during setting, operation and cleaning must be limited to levels that do not have adverse effects on persons.

### 36.11    External radiation

Machinery must be designed and constructed in such a way that external radiation does not interfere with its operation.

## 36.12   Laser radiation

Where laser equipment is used, the following should be taken into account:

- laser equipment on machinery must be designed and constructed in such a way as to prevent any accidental radiation.
- laser equipment on machinery must be protected in such a way that effective radiation, radiation produced by reflection or diffusion and secondary radiation do not damage health.
- optical equipment for the observation or adjustment of laser equipment on machinery must be such that no health risk is created by laser radiation.

## 36.13   Emission of hazardous materials and substances

Machinery must be designed and constructed in such a way that risks of inhalation, ingestion, contact with the skin, eyes and mucous membranes and penetration through the skin of hazardous materials and substances which it produces can be avoided.

Where a hazard cannot be eliminated, the machinery must be so equipped that hazardous materials and substances can be contained, evacuated, precipitated by water spraying, filtered or treated by another equally effective method.

Where the process is not totally enclosed during normal operation of the machinery, the devices for containment and/or evacuation must be situated in such a way as to have the maximum effect.

## 36.14   Risk of being trapped in a machine

Machinery must be designed, constructed or fitted with a means of preventing a person from being enclosed within it or, if that is impossible, with a means of summoning help.

## 36.15   Risk of slipping, tripping or falling

Parts of the machinery where persons are liable to move about or stand must be designed and constructed in such a way as to prevent persons from slipping, tripping or falling on or off these parts.

Where appropriate, these parts must be fitted with handholds that are fixed relative to the user and that enable them to maintain their stability.

# 37.   Maintenance

## 37.1   Machinery maintenance

Adjustment and maintenance points must be located outside danger zones. It must be possible to carry out adjustment, maintenance, repair, cleaning and servicing operations while machinery is at a standstill.

If one or more of the above conditions cannot be satisfied for technical reasons, measures must be taken to ensure that these operations can be carried out safely (see section 1.2.5).

In the case of automated machinery and, where necessary, other machinery, a connecting device for mounting diagnostic fault-finding equipment must be provided.

Automated machinery components which have to be changed frequently must be capable of being removed and replaced easily and safely. Access to the components must enable these tasks to be carried out with the necessary technical means in accordance with a specified operating method.

## 37.2   Access to operating positions and servicing points

Machinery must be designed and constructed in such a way as to allow access in safety to all areas where intervention is necessary during operation, adjustment and maintenance of the machinery.

## 37.3   Isolation of energy sources

Machinery must be fitted with means to isolate it from all energy sources. Such isolators must be clearly identified.

They must be capable of being locked if reconnection could endanger persons. Isolators must also be capable of being locked where an operator is unable, from any of the points to which he has access, to check that the energy is still cut off.

In the case of machinery capable of being plugged into an electricity supply, removal of the plug is sufficient, provided that the operator can check from any of the points to which he has access that the plug remains removed.

After the energy is cut off, it must be possible to dissipate normally any energy remaining or stored in the circuits of the machinery without risk to persons.

As an exception to the requirement laid down in the previous paragraphs, certain circuits may remain connected to their energy sources in order, for example, to hold parts, to protect information, to light interiors, etc. In this case, special steps must be taken to ensure operator safety.

### 37.4    Operator intervention

Machinery must be so designed, constructed and equipped that the need for operator intervention is limited.

If operator intervention cannot be avoided, it must be possible to carry it out easily and safely.

## 37.5    Cleaning of internal parts

The machinery must be designed and constructed in such a way that it is possible to clean internal parts which have contained dangerous substances or preparations without entering them; any necessary unblocking must also be possible from the outside. If it is impossible to avoid entering the machinery, it must be designed and constructed in such a way as to allow cleaning to take place safely.

# 38.　　Marking obligation EU-Guidelines

| Number | Description |
|--------|-------------|
| 2001/95/EC | General product safety |
| 2007/47/EC | Medical Device Directive (MDD) |
| 95/16/EC | Lifts |
| 89/106/EWC | Construction products |
| 97/23/EC | Pressure Equipment Directive (PED) |
| 87/404/EWC | Simple pressure Vessels |
| 06/95/EC | Low Voltage Directive (LVD) |
| 2004/108/EWC | Electromagnetic compatibility (EMC) |
| 93/15/EWC | Explosive for civil uses |
| 99/5/EC | Radio and telecommunications terminal equipment |
| 2009/142/EC | Gas appliance sector |
| 94/9/EC | Equipment and protective systems in potentially explosive atmospheres (ATEX) |
| 2000/14/EC | Noise emission in the environment |
| 98/79/EC | In vitro diagnostic medical devices |
| 2006/42/EC | Machinery Directive |
| 2009/127/EC | Amending Directive with regard to machinery for pesticide application |
| 93/42/EWC | Medical devices |
| 2004/22/EC | Measuring Instruments |
| 2009/23/EC | Non-automatic weighing instruments |
| 2006/95/EWC | Low Voltage Directive |
| 2009/125/EC | Ecodesign Directive |
| 99/36/EC | Transportable pressure equipment |
| 89/686/EWC | Personal protective equipment |
| 2001/65/EC | Internal Market Accounting Directive |
| 96/98/EC | Marine Equipment Directive |
| 2000/9/EC | Cableway installations |
| 88/378/EWC | Toy safety |
| 92/42/EW | Hot-water boilers |

# 39.    Links

EU-Guidelines
www.ce-guidelines.eu

Harmonised Standards
www.ce-guidelines.eu

RAPEX
http://ec.europa.eu/consumers/dyna/rapex/rapex_archives_en.c
fm

Certification QS-Zürich
www.quality-service.ch


DIRECTIVE Leitfaden
www.bmas.de

QS-Engineering
www.qs-engineering.ch

Blueguide
http://ec.europa.eu/enterprise/search/index_de.htm?q=blueguid
e

# 40. Sources

Diverse sections of this book are excerpts of:
- The Machinery Directive 2006/42/EC
- Several ISO DIN EN Standards
- The Blue Guide
- Different regulations
- Different statutory orders

# 41. Translation

Translations of this book in the following languages are in preparation:

- French
- Spanish

# 42.    Index

www.tredition.de

## Über tredition

Der tredition Verlag wurde 2006 in Hamburg gegründet. Seitdem hat tredition Hunderte von Büchern veröffentlicht. Autoren können in wenigen leichten Schritten print-Books, e-Books und audio-Books publizieren. Der Verlag hat das Ziel, die beste und fairste Veröffentlichungsmöglichkeit für Autoren zu bieten.

tredition wurde mit der Erkenntnis gegründet, dass nur etwa jedes 200. bei Verlagen eingereichte Manuskript veröffentlicht wird. Dabei hat jedes Buch seinen Markt, also seine Leser. tredition sorgt dafür, dass für jedes Buch die Leserschaft auch erreicht wird

Autoren können das einzigartige Literatur-Netzwerk von tredition nutzen. Hier bieten zahlreiche Literatur-Partner (das sind Lektoren, Übersetzer, Hörbuchsprecher und Illustratoren) ihre Dienstleistung an, um Manuskripte zu verbessern oder die Vielfalt zu erhöhen. Autoren vereinbaren unabhängig von tredition mit Literatur-Partnern die Konditionen ihrer Zusammenarbeit und können gemeinsam am Erfolg des Buches partizipieren.

Das gesamte Verlagsprogramm von tredition ist bei allen stationären Buchhandlungen und Online-Buchhändlern wie z. B. Amazon erhältlich. e-Books stehen bei den führenden Online-Portalen (z. B. iBook-Store von Apple) zum Verkauf.

Seit 2009 bietet tredition sein Verlagskonzept auch als sogenanntes "White-Label" an. Das bedeutet, dass andere Personen oder Institutionen risikofrei und unkompliziert selbst zum Herausgeber von Büchern und Buchreihen unter eigener Marke werden können.

Mittlerweile zählen zahlreiche renommierte Unternehmen, Zeitschriften-, Zeitungs- und Buchverlage, Universitäten, Forschungseinrichtungen, Unternehmensberatungen zu den Kunden von tredition. Unter www.tredition-corporate.de bietet tredition vielfältige weitere Verlagsleistungen speziell für Geschäftskunden an.

tredition wurde mit mehreren Innovationspreisen ausgezeichnet, u. a. Webfuture Award und Innovationspreis der Buch-Digitale.

tredition ist Mitglied im Börsenverein des Deutschen Buchhandels.

21182340R00093

Printed in Great Britain
by Amazon